NEW from Idea Group Publishing

Qualitative Research in IS: Issues and Trends

Table of Contents

Qualitative Research in IS: Issues and Trends

Eileen M. Trauth
Northeastern University, USA

IDEA GROUP PUBLISHING

Hershey • London • Melbourne • Singapore

Aquisitions Editor: Mehdi Khosrowpour
Managing Editor: Jan Travers
Development Editor: Michele Rossi
Copy Editor: Maria Boyer
Typesetter: Tamara Gillis
Cover Design: Deb Andree
Printed at: Sheridan Books

Published in the United States of America by
        Idea Group Publishing
        1331 E. Chocolate Avenue
        Hershey PA 17033-1117
        Tel: 717-533-8845
        Fax: 717-533-8661
        E-mail: cust@idea-group.com
        Web site: http://www.idea-group.com

and in the United Kingdom by
        Idea Group Publishing
        3 Henrietta Street
        Covent Garden
        London WC2E 8LU
        Tel: 44 20 7240 0856
        Fax: 44 20 7379 3313
        Web site: http://www.eurospan.co.uk

Library of Congress Cataloging-in-Publication Data

Trauth, Eileen Moore.
     Qualitative research in IS : issues and trends / Eileen M. Trauth.
        p. cm.
     Includes bibliographical references and index.
     ISBN 1-930708-06-8 (cloth)
      1. Electronic data processing--Research. 2. Information technology--
     Research. I. Title.

QA76.27 .T75 2001
004--dc21   00-054187

British Cataloguing in Publication Data
A Cataloguing in Publication record for this book is available from the British
Library.

ISSUES FOR THE IS PROFESSION

Preface

As information technologies have evolved, so too has our understanding of the information systems that employ them. A significant part of this evolving understanding is the role of the human contexts within which information systems are situated. This, in turn, has led to the need for appropriate methods of studying information systems in their context of use. There is a growing consensus that qualitative research methods offer important benefits to the study of information systems. This recognition has spawned the demand for more in-depth discussion of the various types of qualitative IS methods so that researchers can determine ones that are most appropriate for addressing their particular research problems. The objective of *Qualitative Research in IS: Issues and Trends* is to address this need. Its intent is to assist IS researchers in their efforts to learn about and employ qualitative methods for IS research.

A significant portion of established and emerging IS researchers are grappling with the issue of learning about new research methods even as they struggle to keep up with new information technologies. This is especially the case for qualitative methods. Many current IS researchers learned research methodology at a time when quantitative analysis was deemed the only legitimate scientific approach. They are now confronted with the need to teach themselves these new qualitative methods. Others may have developed expertise in the use of one particular qualitative method while in school but would now like to learn about alternative qualitative methods. Finally members of the community of qualitative IS researchers need to hear from each other about the challenges and rewards of employing qualitative methods for information systems research.

This book begins by considering trends in the choice of qualitative methods for information systems research. The book then goes on to explore concrete issues that researchers have encountered in the use of a particular qualitative method. It ends by raising issues for the IS profession and suggesting responses. To achieve its objective, this book draws upon the collective expertise of distinguished scholars who employ qualitative methods in their own research. These individuals are widely known for their experience in conducting qualitative research. They include: Richard Baskerville, Dubravka Cecez-Kecmanovic, Heinz Klein, Allen Lee, Enid Mumford, Michael Myers, Steve Sawyer, Ulrike Schultze, Cathy Urquhart and Eleanor Wynn. These authors address issues for individual researchers and for the IS profession from the vantage point of their own experiences. They draw upon their own published work to consider issues that they have encountered and the ways in which they have resolved them.

The audience for this book includes both IS students and IS academics. This book is suitable for use in a doctoral seminar in qualitative research in IS. It will also find an audience among established researchers. Anyone currently engaged in conducting IS research who would like to learn more about employing qualitative methods will be interested in this book.

This book fills a unique space in the IS research literature. Unlike journal articles, which focus on research findings, the focus of these chapters is on the methodology issues themselves. The details of specific research projects provide the backdrop for the discussion of methodological issues. The book also complements existing books on qualitative methodology. Some books address qualitative methodology in general. The subject of the research is not information systems per se, but rather social science in general. A problem that IS researchers often encounter with using such books is the translation of the general ideas into the specific context of IS research. This book is different in that it shows applications of these qualitative methods in a context relevant to IS researchers. Other books are about a single IS research project that employs qualitative methodology. This book is different from these books because it provides the reader with more than one example of the application of qualitative methods to IS research. A third category of books contains collections of chapters that focus on research methodology as applied to information systems research. This book falls into this category. Like other books in this category, it is a collection of chapters by a variety of established authors. However, what makes it distinctive is that the focus of each chapter is methodology rather than the research findings. While the authors discuss the content of their research projects, methodological issues are in the foreground.

It has been a privilege serving as the editor of this volume. But as every editor knows, there are those people without whose contribution a book would never exist. First, I would like to thank Idea Group Publishing for recognizing the need for this book and for asking me to edit it. Second, I would like to thank the contributing authors who agreed to participate in this project. Their willingness to discuss their own experiences, issues and challenges has made this book what it is. Third, I would like to thank those individuals who have helped to turn the manuscript into the final product. I would like to thank my research assistant Andrew Esposito for his efforts under tight deadlines. Last, but not least, I would like to thank the editorial staff at Idea Group Publishing whose patience and assistance throughout this project has made the work of editing this book an enjoyable and rewarding experience.

Eileen Trauth
December 2000

Trends
in the
Choice of
Qualitative
Methods

Chapter I

The Choice of Qualitative Methods in IS Research

Eileen M. Trauth
Northeastern University, USA

INTRODUCTION

In this introductory chapter I set the stage for the remaining chapters by discussing factors that influence the choice of qualitative methods for information systems research. In doing so, I provide examples from my own work as well as that of other qualitative researchers in the IS field. I consider these influencing factors in order to highlight the interplay between methodological choices and the context within which they occur. Just as decisions about information systems need to be considered within their contexts of use, so too do choices about qualitative methods for information systems research.

In successive waves throughout my career, I have broadened the scope of the qualitative research methods I have chosen to use. In doing so, I have also expanded the range of issues I have had to confront. My qualitative research initially took the forms of case study and policy analysis as I followed the telecommunications privatization movement in the U.S. (Trauth 1979, 1986; Trauth, et al., 1983, 1991). The next stage of my journey began in 1989 when I was developing a research plan for a country-level case study of Ireland's emerging information economy. In this project the scope (the entire country), the level of analysis (societal), and the interpretive nature of the research presented significant challenges. First, countrywide case studies are typically statistical studies.[1] Second, the organizational level of analysis is typical for published qualitative IS research. Third, this research project represented a shift from the more positivist use of qualitative methods in my policy analysis research to the interpretive use of qualitative methods for theory development.[2] For all these reasons I had little by way of exemplars in the IS field to guide me through the morass of methodological choices. After I chose an ethnographic approach, I was forced to extrapolate, as best I could, from examples in other fields.[3]

A new phase of my involvement with qualitative research methods began in 1997 when I was confronted with two new issues. One was adapting interpretive research methods for the virtual realm. I was particularly interested in the process of applying interpretive methods such as ethnography — which assumed both face-to-face data collection and extended periods of time in the field interacting with the research subjects — to study the behavior of virtual groups. This research led to the other issue: developing and assessing interpretive research methodologies. Whereas positivist research can appeal to established statistical tests to certify reliability and validity, interpretive information systems (IS) research has not had such a tradition. Until quite recently, there has been little available in the IS literature to guide the interpretive researcher. For these reasons I became interested in contributing to the development of a cumulative body of knowledge regarding the use and assessment of interpretive research.[4]

Throughout all these phases of my research career what I most often sought were examples to help show me the way. I wanted to see how others were engaging with research issues that were similar to mine, whether they were about the choice of appropriate method, the particulars of data collection and analysis associated with a given method, or finding appropriate evaluative criteria once the method was chosen. Unfortunately, I was often frustrated in my attempts to do so. The public discussion of qualitative methods in information systems research is fairly recent and heavily influenced by geography. The "Manchester Conference" on information systems research is generally viewed as initiating the discussion (Mumford et al., 1985). Another European conference held in Copenhagen in 1990 (Nissen et al., 1991) continued the discussion of qualitative methods and deepened the consideration of specific methods. Surrounding these two conferences and the books that resulted from them were occasional journal articles that either discussed or employed qualitative methods (e.g., Benbasat et al., 1987; Kaplan and Duchon, 1988; Lee, 1989; Markus, 1983).

But it was not until the 1990s that qualitative research was consistently published in the major IS journals in which Americans publish predominantly.[5] The focus of those articles that have been published has understandably been on the research, itself, rather than on the mechanics of methodology. Throughout this journey as I sought information and answers, I encountered a growing community of people who were asking similar questions.

For these reasons I was delighted to be asked by Idea Group Publishing to edit a book on qualitative methods in information systems research. Editing this book gives me an opportunity to help address an important need in the IS research community. The purpose of this book is to facilitate discussion of the actual use of qualitative research methods. There are two audiences for the book. One audience consists of students who are learning about qualitative IS research methods in a structured setting. These readers would typically be graduate students in a research methods course or Ph.D. students working on their dissertations under the tutelage of an advisor.

The other audience comprises independent learners. These might be Ph.D. students in a research degree that has no formal classes on qualitative research methods. Or they might be academics who want to learn more about the use of qualitative methods. Such people may not have studied qualitative methods in graduate school but now find themselves in a position of wanting to employ them. They might be mid-career academics, trained in positivist and quantitative methods, who are trying to reach beyond their current comfort level by teaching themselves new research methods. Finally, they may be qualitative researchers who endeavor to expand their methodological horizons by learning more about other qualitative methods that they have not used. These IS researchers already engage in qualitative research but want to add depth to their understanding about research methodology.

The remainder of this chapter lays the foundation for the book by considering five factors which can influence the choice of qualitative methods for information systems research. These factors are then illustrated in the ensuing chapters which highlight the range of issues that can arise in the use of qualitative methods for IS research. Some

chapters consider specific issues associated with a particular methodological choice. Other chapters consider challenges for the IS profession as a whole. Along the way, the reader can accompany the authors
as they trace through the whys and wherefores of employing the various
qualitative methods. In this way, the chapters serve two informational
goals: they illustrate the use of particular qualitative research methods
while they critically analyze issues associated with doing so.

FACTORS INFLUENCING THE CHOICE
OF QUALITATIVE METHODS IN IS RESEARCH

In reflecting upon my own decisions in order to discuss the choice
of qualitative methods and the factors that can influence those choices,
I draw upon two research projects in which I chose qualitative methods.
Although both of these projects involved interpretive research methods, the characteristics of these projects are sufficiently different in that
I have encountered and had to cope with different issues in each of them.
One project is an exploration of sociocultural influences on Ireland's
information economy and the subsequent sociocultural impacts that
have resulted (Trauth, 1993, 1995, 1996, 1999, 2000b; Trauth and Pitt,
1992). I have been the sole researcher in this multi-year ethnographic
study. Through interviews, participant observation and document
analysis, I collected data at both micro (IT organizations) and macro
(Irish society) levels. The other project was a collaborative project that
explored the information exchange among individuals employing
group support system technology to discuss a high threat topic (Trauth
and Jessup, 2000). Our approach was to analyze the data by using two
different epistemological lenses: positivist and interpretive. Using these
examples and others from the literature, I will now consider five influences
on the choice of qualitative methods in IS research.

The Research Problem

Some would argue that the nature of the research problem should
be the most significant influence on the choice of a research methodology. That is, *what* one wants to learn determines *how* one should go
about learning it. The qualitative methods literature is replete with this
rationale for the choices that have been made.

Field studies have been used in a variety of settings to uncover
subtleties of process and impact related to the use of information

technology. Heaton (1998) chose the interpretive methods of observation, interview and document analysis to examine the social construction of computer-supported cooperative work in two different cultures. Her choice was influenced by her desire to examine what "culture" meant to her informants and how they reflected that meaning in the way they designed systems. Sayer (1998) adopted a postmodern ethnographic approach to reflect the organizational transformation process that accompanies the implementation of business process re-engineering. Komito (1998) turned to ethnography to enable him to highlight the limitations of planned electronic communication systems to replicate existing work practices. To inform their design of an interface to an air traffic control database, Bentley et al. (1992) chose ethnographic methods to study the work practices of air traffic controllers. Finally, Walsham and Sahay (1999) conducted extensive interviews to gain an in-depth knowledge from stakeholders regarding the implementation of geographical information systems for real district-level administrative applications in India.

Another methodological choice in the IS literature that reflects the influence of the research problem is document analysis. For example, Phillips (1998) employed public discourse analysis to reveal the way in which concerns about anonymity, surveillance, security and privacy are integrated into public understanding of a consumer payment system. Davidson (1997) employed narrative analysis to analyze project history narratives contained in research interviews. And Ang and Endeshaw (1997) drew from legal case analysis to develop an approach for representing prototypical disputes in IT management.

In my country-case study of Ireland's information economy, I came to the choice of ethnographic methods because of my desire to uncover the "story behind the statistics" about Ireland's information sector. My research problem was exploring the role of sociocultural context in the development of a nation's information economy. I wanted to identify the influence of culture, history, public policy and other societal factors on an emerging information sector. I determined that the best way to obtain the information I sought was to immerse myself in the world in which they were occurring. I decided to observe people and events, analyze documents and literature, and talk to people both formally and informally. In order to do this, I needed to spend an extended period of time in the field.

The Researcher's Theoretical Lens

In the debate about alternatives to positivist research, some have suggested that the choice of method may not be a choice at all. They would argue that methods are adopted in conformance with the episte-mological orthodoxy of positivism (Van Maanen, 1979) or in reaction to it (Markus, 1997). Whether one agrees with this viewpoint or not, it is clear that another important influence on the choice of research method is the theoretical lens that is used to frame the investigation. Throughout the literature about qualitative approaches to IS research, there have been papers that explore this influence. A good example is Orlikowski and Baroudi's (1991) examination of these theoretical lenses and how they have shaped IS research. Citing Chua's (1986) classification of research epistemologies, they go on to describe these three lenses through which IS research is conducted and the influences of these lenses on the choice of research method.

Positivist studies are premised on the existence of a priori fixed relationships within phenomena which are typically investigated with structured instrumentation. Such studies are primarily to test theory... (p. 5)

Interpretive studies assume that people create and associate their own subjective and intersubjective meanings as they interact with the world around them... [T]he intent is to understand the deeper structure of a phenomenon ... to in-crease understanding of the phenomenon within cultural and contextual situations... (p. 5)

Critical studies aim to critique the status quo, through the exposure of what are believed to be deep-seated, structural contradictions within social systems, and thereby to transform these alienating and restrictive social conditions. (pp. 5-6)

Given the dominant position of positivism in IS research it is not surprising that some of the qualitative work in North America has been in the positivist tradition (Eisenhardt, 1989; Lee, 1989; Markus, 1983; Paré and Elam, 1997) or that it has attempted to bridge the positivist/ quantitative - interpretive/qualitative divide (Gallivan, 1997; Kaplan and Duchon, 1988; Lee, 1991).

But interpretivism is the lens most frequently influencing the choice of qualitative methods. This is because of the assumption that "our knowledge of reality is a social construction by human actors" hence, objective, value-free data cannot be obtained (Walsham, 1995a, p. 376). In addition to ethnographic methods, the interpretive epistemology has also spawned IS research employing hermeneutic methods (e.g., Boland, 1985, 1991; Lee, 1994; Trauth and Jessup, 2000).

Myers' paper on critical ethnography (1997) helps to bridge the understanding gap between interpretive and critical research — the alternatives to the dominant lens of positivism. Ngwenyama and Lee (1997) use the critical lens to guide their approach to examining information richness theory. Doolin (1998) argues that a research approach based on critical theory is needed in order to view information technology within a broader context of social and political relations.

In our study of information exchanges among individuals employing group support systems technology, the influence of theoretical lens was felt on both the choice of research method and on the research findings. In our study, the positivist lens and the quantitative analysis told us *that* people communicated. But it could not tell us *what* they communicated or *why* they communicated as they did. It could not provide us with an in-depth look at the worldviews that sat behind the "facts" shared by the participants. Nor could it provide us with the reasons behind their behavior. The interpretive lens and qualitative analysis of the texts were needed to get at the *why* of the information sharing behavior and the mechanics of *how* within that particular context.[6]

The Degree of Uncertainty Surrounding the Phenomenon

The amount of uncertainly surrounding the phenomenon under study is another important factor in the choice of qualitative research methods. From a positivist perspective, the less that is known about a phenomenon the more difficult it is to measure it. For example, Benbasat et al. (1987) explain that the case study approach is appropriate for IS research areas in which few previous studies have been carried out. Paré and Elam (1997) employed a positivist epistemology to consider how to build theories of IT implementation using case study methods.

From an interpretive point of view, Orlikowski (1993) explains her choice of qualitative methods to study the adoption of CASE tools. She chose grounded theory because there had been no systematic examination of the organizational changes accompanying the introduction of CASE tools. Hence, no change theory of CASE tools adoption and use had been established. Galliers and Land (1987) point to the added complexity that comes from a view of information systems that includes their relations with people and organizations. Accompanying this broadened scope of study comes greater imprecision and the potential for multiple interpretations of the same phenomenon. Under these circumstances alternatives to quantitative measurement are needed.

For example, in my study of Ireland's information economy, there was considerable uncertainty regarding which sociocultural factors were relevant to Ireland's nascent information economy. Further, I did not know what kind of influence they were exerting on it. In addition, the societal impact of that information economy was barely in evidence when the research began. A questionnaire was not practical since I was unable to pre-specify relevant sociocultural variables. In addition, a survey with its fixed questions and categories would not have given me the flexibility that I believed was needed in such a dynamic research setting. The sociocultural influences were in a constant state of flux, were happening all around me, and the most significant impacts of Ireland's decision to develop an information sector did not begin to emerge until the research was well underway.[7] This uncertainty influenced me to choose ethnography and grounded theory.

The Researcher's Skills

As I mentioned in the Introduction, an individual's level of skill, knowledge and experience in using qualitative research methods is a significant influence when deciding whether or not to employ them in IS research. To the extent that qualitative research methods are part of one's methodology portfolio, an individual can select from them when appropriate. But an individual is less likely to choose to employ these methods if he or she has not learned about or does not have experience with qualitative methods. A researcher's skill in using qualitative methods is an influencing factor not only during the dissertation but also throughout her or his career.

At a 1997 ICIS conference panel discussion on qualitative research methods the senior faculty, junior faculty and graduate student perspectives all spoke to the challenge of developing and maintaining appropriate skills with qualitative methods. Boudreau (1997) represented the new breed of IS doctoral students for whom qualitative research methods is considered a viable alternative. For her, the educational need was to have more exemplars. Representing the junior faculty perspective Kaarst-Brown (1997) expressed the educational challenge for those who studied qualitative methods in graduate school and perhaps used them in their research. Not only are there ever-changing technologies to learn about, but those who choose qualitative research methods are taking on the challenge of keeping abreast of the new methodologies and new interpretations of existing methodologies.

Representing the senior faculty perspective (Trauth, 1997b), I addressed the educational issues for those individuals who received their education and perhaps tenure at a time when qualitative methods were not a choice for information systems research. Individuals in this position are now confronting phenomena and research issues for which their training has not prepared them. Their training in quantitative research methods, alone, did not equip them with the complete repertoire of research tools needed to study the organizational, societal and cross-cultural issues accompanying the widespread deployment of information technology in the twenty-first century. For these individuals the lack of skill and experience with qualitative methods may well function as a barrier to employing this research approach.

In discussing the role of research training in the United States, Orlikowski (1991) points to institutional conditions that have inhibited the teaching of alternative, qualitative research methods such as action research, critical research and interpretive research. She found the cause of this lack of training in qualitative research methods in the functionalist and positivist perspective of American business schools where information systems is typically taught. In discussing institutional conditions within which doctoral studies are conducted and dissertations are written, she suggested that these conditions have inhibited the use of alternative research paradigms and methodologies with long-term implications for the kind of research we might expect in the IS field.

The importance of this institutional influence is reinforced in Schultze's chapter in which she reflects on her decision to choose interpretive methods for her dissertation.[8] Because she was at one of the few places with faculty expertise in interpretive methods, she felt she had methodological opportunities that many others did not have. That the situation is changing is evident in the growing number of qualitative studies published in mainstream IS journals. This reflects both an increase in the number of doctoral students employing such methods and the changing attitude about the acceptance of these methods.

But the change in Ph.D. programs does little to help those who are beyond that stage of their careers. These individuals must develop skills with qualitative methods by other means. In response to this need, books have been written (Lee et al., 1997; Mumford et al., 1985; Nissen et al., 1991), journals that actively encourage qualitative research have been established,[9] special issues of IS journals have been devoted to articles employing qualitative methods[10], and the "Qualitative Research in Information" section of ISWorld was developed.[11] Through the growing body of literature that illustrates and explains various qualitative methods for the conduct of IS research, IS scholars are helping to enhance researchers' skill in the use qualitative methods.[12] Finally the peer review process of journals is being used to help researchers increase their understanding of qualitative research methods. In the case of our positivist/interpretive GSS study, the role of the review process in helping us to develop our work is documented in Lee (1999).

Academic Politics

The final factor influencing the choice of qualitative methods for information systems research that I will discuss relates to the norms and values of the IS field, the institution at which one works and the status that one holds there, and the country in which that institution is located. These all serve to influence the choice of qualitative methods for IS research. Factors such as the country in which one works, one's status in the profession — whether one has completed the Ph.D., whether one has a tenured position, and one's academic rank — and the particular inclinations of the university at which one works all influence the choice of research methods.

The norms and values of the field are expressed in a variety of ways, and are reinforced during one's education and beyond. What is taught in research methods seminars sets the standard for "acceptable" research. Later, through advice to junior faculty, peer review of journal articles and the tenure review process, adherence to those norms and values is enforced. In a very real sense, then, journal editors and reviewers, department chairs, and others in positions of authority in the IS field serve as methodological gatekeepers.

In a review of predominately American IS research in the 1980s Orlikowski and Baroudi (1991) argued that information systems research was being guided by a dominant, positivist worldview that sanctioned methodologies consistent with a positivist research perspective. Fitzgerald and Howcroft (1998) tell two compelling "tales" to illustrate the polarization of positions into "hard" and "soft" perspectives. But their contribution to understanding this factor goes beyond the clarity they bring to the positions; they also offer suggestions for bridging the gap between the two extreme positions. Klein and Myers (1999) also contribute to closing the epistemological divide. By developing a set of principles for conducting and evaluating interpretive field studies, they offer a concrete response to positivists' complaint that "anything goes" in interpretive research.

ORGANIZATION OF THE BOOK

In the previous section I considered some of the factors influencing the choice of qualitative methods for IS research. The issues that accompany this choice are at the core of this book. These issues arise at both the individual and the professional levels within the IS field. Eleanor Wynn provides a textured background for the consideration of specific issues by discussing past and future trends in the use of qualitative methods. She writes from her experiences not only in conducting qualitative research, but also from her experiences as an editor of a leading IS journal that publishes qualitative IS research.[13] Like all the chapters in this book, this chapter employs examples from actual research projects to illustrate the trends. This chapter provides the reader with a sense of the whole about the implications of the choice of qualitative research methods for information systems research.

The remaining chapters in the book focus on issues related to the actual use of specific qualitative research methods. The discussion of issues that result from the choice of qualitative research methods is presented on two levels. Chapters by Enid Mumford, Ulrike Schultze, Cathy Urquhart, Dubravka Cecez-Kecmanovic, and Steve Sawyer address issues at the individual level of analysis. Drawing upon their own experiences in using a particular method in a particular project, they discuss concrete issues associated with the choice of a particular research methodology.

The authors provide examples from actual research projects they have conducted in order to illustrate the issues at hand. The objective is to have the reader come away with a general understanding of what the method is, why someone would want to use it, and issues that are associated with using it. Overall, the reader gets a good sense of what it would be like to employ the method being discussed.

To facilitate learning, each chapter has a consistent format. Following an introduction to the particular research method, the chapter raises issues within the context of a particular research study that has employed that particular method. These are issues encountered in the conduct of the research or in the publication of the results. It is expected that an independent learner could read the chapter and develop a better understanding about the *actual use* of the method; participants in a seminar course could use this chapter as the basis for a rich class discussion.

The three chapters by Richard Baskerville, Heinz Klein and Michael Myers, and Allen Lee address issues associated with the choice of qualitative methods at the level of the IS profession. Baskerville examines a particular methodology from the point of view of professional risks associated with choosing it. In doing so, he also raises some issues that are applicable to qualitative research in general. Klein and Myers address the issue of developing a cumulative tradition in qualitative IS research. Through this cumulative tradition, they argue, qualitative IS research will more easily deepen and diversify. Lee draws upon his own research as well as his experience as an IS journal editor[14] in looking toward the future. He presents challenges for the IS research community that suggest future trends in qualitative IS research.

In the final chapter of the book, I draw from the issues and trends that are presented in this book to generate lessons learned about using qualitative methods in the conduct of information systems research. These lessons relate both to the choice of a given method and to the development of the IS field. Consequently, these lessons should be heard by both individual IS researchers and the IS profession.

CONCLUSION

In this chapter I explored the rationale for producing this book, and explained what motivated it and why it is being written. I also suggested factors that are influencing the choice of qualitative methods. Finally,

I explained the organization of this book. The acceptance of qualitative methods for the conduct of IS research — particularly in the United States — is a product of the 1990s. Although qualitative scholars can still encounter difficulty in getting their work published in certain journals, there is a growing consensus that qualitative methods constitute a much needed approach to the study of information systems. The problem is that there have not been enough examples of research employing qualitative methods. In fashioning a research design, researchers want to look to examples that are similar to their own. In addition, there is a need for more in-depth discussion of the various types of qualitative IS research so that researchers can determine which methods are most appropriate to their particular research questions. It is for these reasons that this book has been written.

REFERENCES

Ang, S. and Endeshaw, A. (1997). Legal Case Analysis in IS Research: Failures in Employing and Outsourcing for IT Professionals. In A.S. Lee, J. Liebenau, and J.I. DeGross (Eds.), *Information Systems and Qualitative Research* (pp. 497-523). London: Chapman & Hall.

Benbasat, I., Goldstein, D.K. and Mead, M. (1987). The Case Research Strategy in Studies of Information Systems. *MIS Quarterly,* 11(3), 369-386.

Bentley, R., Hughes, J.A., Randall, D., Rodden, T., Sawyer, P., Shapiro, D. and Sommeville, I. (1992). Ethnographically-informed Systems Design for Air Traffic Control. In J. Turner and R. Kraut (Eds.), *Sharing Perspectives: Proceedings of ACM Conference on Computer-supported Cooperative Work* (pp.123-129). New York: ACM Press.

Boland, R.J., Jr. (1991). Information System Use as a Hermeneutic Process. In H.-E. Nissen, H.K. Klein and R. Hirschheim (Eds.), *Information Systems Research: Contemporary Approaches & Emergent Traditions* (pp. 439-458). Amsterdam: North-Holland.

Boland, R.J. (1985). Phenomenology: A Preferred Approach to Research on Information Systems. In Mumford, E., Hirschheim, R.A., Fitzgerald, G. and WoodHarper, T. (Eds.), *Research Methods in Information Systems* (pp.193-201). Amsterdam: NorthHolland.

Boudreau, M-C. (1997). A Report from 8.2 to ICIS: Observations and Lessons on Qualitative Research in IS from the June 1997 Conference of IFIP Working Group 8.2 — The Newcomer's View. Panel

Presentation at the International Conference on Information Systems, Atlanta, GA, December.

Butler, T. (1998). Towards a Hermeneutic Method for Interpretive Research in Information Systems. *Journal of Information Technology*, Special Issue on Interpretive Research in Information Systems, 13(4), 285-300.

Chua, W.F. (1986). Radical Developments in Accounting Thought. *The Accounting Review*, 61, 601-632.

Davidson, E.J. (1997). Examining Project History Narratives: An Analytic Approach. In A.S. Lee, J. Liebenau, and J.I. DeGross (Eds.), *Information Systems and Qualitative Research* (pp. 123-148). London: Chapman & Hall.

Doolin, B. (1998). Information Technology as Disciplinary Technology: Being Critical in Interpretive Research on Information Systems. *Journal of Information Technology*, Special Issue on Interpretive Research in Information Systems, 13(4), 301-312.

Eisenhardt, K.M. (1989). Building Theories from Case Study Research. *MIS Quarterly*, 14(4), 532-550.

Fitzgerald, B. and Howcroft, D. (1998). Towards Dissolution of the IS Research Debate: From Polarization to Polarity. *Journal of Information Technology*, Special Issue on Interpretive Research in Information Systems, 13(4), 313-326.

Fransman, M. (1995). *Japan's Computer and Communications Industry*. Oxford: Oxford University Press.

Galliers, R.D. and Land, F.F. (1987). Choosing Appropriate Information Systems Research Strategies. *Communications of the ACM*, 30(11), 900-902.

Gallivan, M.J. (1997). Value in Triangulation: A Comparison of Two Approaches for Combining Qualitative and Quantitative Methods. In A.S. Lee, J. Liebenau, and J.I. DeGross (Eds.), *Information Systems and Qualitative Research* (pp. 417-443). London: Chapman & Hall.

Gopal, A. and Prasad, P. (2000). Understanding GDSS in Symbolic Context: Shifting the Focus from Technology to Interaction. *MIS Quarterly*, Special Issue on Intensive Research, 24(3), 509- 546.

Heaton, L. (1998). Talking Heads vs. Virtual Workspaces: A Comparison of Design Across Cultures. *Journal of Information Technology*, Special Issue on Interpretive Research in Information Systems, 13(4), 259-272.

Heeks, R. (1996). *India's Software Industry.* New Delhi: Sage.

Kaarst-Brown, M.L. (1997). A Report from 8.2 to ICIS: Observations and Lessons on Qualitative Research in IS from the June 1997 Conference of IFIP Working Group 8.2 — My View as an Assistant Professor. Panel Presentation at the International Conference on Information Systems, Atlanta, GA, December.

Kaplan, B. and Duchon, D. (1988). Combining Qualitative and Quantitative Methods in Information Systems Research: A Case Study. *MIS Quarterly,* 12(4), 571-586.

Kelly, T. (1987). *The British Computer Industry.* London: Croom Helm.

Klein, H.K. and Myers, M.D. (1999). A Set of Principles for Conducting and Evaluating Interpretive Field Studies in Information Systems. *MIS Quarterly,* Special Issue on Intensive Research, 23(1), 67-93.

Komito, L. (1998). Paper 'Work' and Electronic Files: Defending Professional Practice. *Journal of Information Technology,* Special Issue on Interpretive Research in Information Systems, 13(4), 235-246.

Lee, A.S. (1999). Editors Comments: The Role of Information Technology in Reviewing and Publishing Manuscripts at *MIS Quarterly. MIS Quarterly,* 23(4), iv-ix.

Lee, A.S. (1994). Electronic Mail as a Medium for Rich Communication: An Empirical Investigation Using Hermeneutic Interpretation. *MIS Quarterly,* 18(2), 143-157.

Lee, A.S. (1991). Integrating Positivist and Interpretive Approaches to Organizational Research. *Organization Science,* 2(4), 342-365.

Lee, A.S. (1989). A Scientific Methodology for MIS Case Studies. *MIS Quarterly,* 13(1), 33-50.

Lee, A.S., Liebenau, J. and DeGross, J.I. (Eds.). (1997). *Information Systems and Qualitative Research.* London: Chapman & Hall.

Luzio, E. (1996). *The Microcomputer Industry in Brazil.* Westport, CT: Praeger.

Markus, M.L. (1983). Power, Politics, and MIS Implementation. *Communications of the ACM,* 26(6), 430-444.

Markus, M.L. (1997). The Qualitative Difference in Information Systems Research and Practice. In A.S. Lee, J. Liebenau, and J.I. DeGross (Eds.), *Information Systems and Qualitative Research* (pp. 11-27). London: Chapman & Hall.

Mumford, E., Hirschheim, R.A., Fitzgerald, G. and WoodHarper, T. (Eds.). (1985). *Research Methods in Information Systems.* Amsterdam: North-Holland.

Myers, M.D. (1997). Critical Ethnography in Information Systems. In A.S. Lee, J. Liebenau, and J.I. DeGross (Eds.), *Information Systems and Qualitative Research,* (pp. 276-300), London: Chapman & Hall.

Nelson, K.M., Nadkarni, S., Narayanan, V.K., Ghods, M. (2000). Understanding Software Operations Support Exerptise: A Revealed Causal Mapping Approach. *MIS Quarterly,* Special Issue on Intensive Research, 24(3), 475-507.

Ngwenyama, O.K. and Lee, A.S. (1997). Communication Richness in Electronic Mail: Critical Social Theory and the Contextuality of Meaning. *MIS Quarterly,* 21(2), 145-167.

Nissen, H.-E., Klein, H.K. and Hirschheim R. (Eds.). (1991). *Information Systems Research: Contemporary Approaches and Emergent Traditions.* Amsterdam: North-Holland.

Orlikowski, W.J. (1993). CASE Tools as Organizational Change: Investigating Incremental and Radical Changes in Systems Development. *MIS Quarterly,* 17(3), 309-340.

Orlikowski, W.J. (1991). Relevance versus Rigor in Information Systems Research: An Issue of Quality — The Role of Institutions in Creating Research Norms, Panel Presentation at the IFIP TC8/WG 8.2 Working Conference on the Information Systems Research Challenges, Perceptions and Alternative Approaches, Copenhagen, Denmark.

Orlikowski, W.J. and Baroudi, J.J. (1991). Studying Information Technology in Organizations: Research Approaches and Assumptions. *Information Systems Research,* 2(1), 1-28.

Paré, G. and Elam, J.J. (1997). Using Case Study Research to Build Theories of IT Implementation. In A.S. Lee, J. Liebenau, and J.I. DeGross (Eds.), *Information Systems and Qualitative Research* (pp. 542-568). London: Chapman & Hall.

Phillips, D. (1998). The Social Construction of a Secure, Anonymous Electronic Payment System: Frame Alignment and Mobilization Around Ecash. *Journal of Information Technology,* Special Issue on Interpretive Research in Information Systems, 13(4), 273-284.

Sayer, K. (1998). Denying the Technology: Middle Management Resistance in Business Process Re-engineering. *Journal of Information Technology,* Special Issue on Interpretive Research in Information Systems, 13(4), 247-258.

Schultze, U. (2000). A Confessional Account of an Ethnography about Knowledge Work. *MIS Quarterly,* Special Issue on Intensive Research, 24(1), 1-39.

Trauth, E.M. (1997a). Achieving the Research Goal with Qualitative Methods: Lessons Learned along the Way. In A.S. Lee, J. Liebenau, and J.I. DeGross (Eds.), *Information Systems and Qualitative Research* (pp. 225-245). London: Chapman & Hall.

Trauth, E.M. (1979). *An Adaptive Model of Information Policy.* Ann Arbor, MI: University Microfilms.

Trauth, E.M. (2000b). *The Culture of an Information Economy: Influences and Impacts in the Republic of Ireland.* Dordrecht, The Netherlands: Kluwer Academic Publishers.

Trauth, E.M. (1993). Educating Information Technology Professionals for Work in Ireland: An Emerging Post-industrial Country. In M. Khosrowpour and K. Loch (Eds.), *Global Information Technology Education: Issues and Trends* (pp. 205-233). Harrisburg, PA: Idea Group Publishing.

Trauth, E.M. (1996). Impact of an Imported IT Sector: Lessons from Ireland. In E.M. Roche and M.J. Blaine (Eds.), *Information Technology Development and Policy: Theoretical Perspectives and Practical Challenges* (pp. 245-261). Aldershot, UK: Avebury Publishing Ltd.

Trauth, E.M. (1986). An Integrative Approach to Information Policy Research. *Telecommunications Policy,* 10(1), 41-50.

Trauth, E.M. (1999). Leapfrogging an IT Labor Force: Multinational and Indigenous Perspectives. *Journal of Global Information Management,* 7(2), 22-32.

Trauth, E.M. 1997(b). A Report from 8.2 to ICIS: Observations and Lessons on Qualitative Research in IS from the June 1997 Conference of IFIP Working Group 8.2 — A Senior Perspective. Panel Presentation at the International Conference on Information Systems, Atlanta, GA December.

Trauth, E.M. (1995). Women in Ireland's Information Industry: Voices from Inside. *Eire-Ireland,* 30(3), 133-150.

Trauth, E.M., Huntley, J., and Pitt, D.C. (1991). The Implementation Game: Deregulating U.S. Telecommunications. In T. Younis (Ed.), *Policy Implementation in Public Administration,* (pp. 103-113), Aldershot, UK: Dartmouth Publishing Co., Ltd.

Trauth, E.M. and Jessup, L. (2000). Understanding Computer-mediated Discussions: Positivist and Interpretive Analyses of Group Support System Use. *MIS Quarterly*, Special Issue on Intensive Research, 24(1), 43-79.

Trauth, E.M. and O'Connor, B. (1991). A Study of the Interaction Between Information Technology and Society: An Illustration of Combined Qualitative Research Methods. In H.-E. Nissen, H.K. Klein and R. Hirschheim (Eds.), *Information Systems Research: Contemporary Approaches & Emergent Traditions* (pp. 131-144). Amsterdam: North-Holland.

Trauth, E.M. and Pitt, D.C. (1992). Competition in the Telecommunications Industry: A New Global Paradigm and Its Limits. *Journal of Information Technology*, 7(1), 3-11.

Trauth, E.M., Trauth, D.M., and Huffman, J.L. (1983). Impact of Deregulation on Marketplace Diversity in the USA. *Telecommunications Policy*, 7(2), 111-120.

Van Maanen, J. (1979). Reclaiming Qualitative Methods for Organizational Research: A Preface. *Administrative Science Quarterly*, 24, 520-526.

Walsham, G. (1995a). The Emergence of Interpretivism in IS Research. *Information Systems Research*, 6(4), 376-394.

Walsham, G. (1995b). Interpretive Case Studies in IS Research: Nature and Method. *European Journal of Information Systems*, 4, 74-81.

Walsham, G. and Sahay, S. (1998). GIS for District-Level Administration in India: Problems and Opportunities. *MIS Quarterly*, Special Issue on Intensive Research, 23(1), 39-65.

ACKNOWLEDGMENTS

This chapter has benefited from the insightful comments and suggestions by Yang Lee, Steve Sawyer and Ulrike Schultze.

ENDNOTES

1 See, for example, Fransman (1995), Heeks (1996), Kelly (1987) and Luzio (1996).

2 It is important to point out that "qualitative" and "interpretive" are not equivalent terms. As will be illustrated in this chapter, qualitative methods can be used not only for interpretive and

critical research but also for positivist research.

3   The examples most commonly used in the methodology books I consulted came from the fields of sociology and education.

4   The purpose of Trauth (1997a) was to contribute to this effort.

5   While it is understood that this book has an international audience, the reflexive style in which this chapter is written causes me to speak primarily about the status of qualitative research methods in the United States, the country in which I received my education and where I engage in my profession.

6   Gopal and Prasad (2000) had similar results.

7   These methodological considerations are discussed in Trauth (2000b) and Trauth and O'Connor (1991).

8   See also Schultze (2000) for further discussion of this research project.

9   The two most prominent IS journals publishing qualitative research are *Information and Organization* (formerly *Accounting, Management and Information Technologies*) and *Information Technology and People*.

10   Two recent special issues of journals that have focused on qualitative methods for information systems research are: *Journal of Information Technology*, Volume 13, Number 4 (December 1998), Special Issue on Interpretive Research in Information Systems; and *MIS Quarterly* Volume 23, Number 1 (March 1999), Volume 24, Number 1 (March 2000), and Volume 24, Number 3 (September 2000), Special Issue on Intensive Research.

11   It is located at: www.auckland.ac.nz/msis/isworld.

12   Examples are Walsham's (1995b) paper on interpretive case studies, Butler's (1998) paper on hermeneutic research in information systems, and Nelson et al.'s (2000) paper on revealed causal mapping methodology.

13   At the time of this writing, Wynn is editor-in-chief of *Information Technology and People*.

14   At the time of this writing, Lee is editor-in-chief of *MIS Quarterly*.

Chapter II

Möbius Transitions in the Dilemma of Legitimacy

Eleanor Wynn
Intel Corporation, USA

INTRODUCTION

Over the past 10 years, qualitative research has gained acceptance in the academic Information Systems (IS) discipline in the U.S., following the lead of European nations and Australia (Lee & Liebenau, 1997). In many leading-edge organizations, principles derived from qualitative research now inform information technology (IT) policies and principles. For instance the idea of communities of practice, which draws on the phenomenological sociology of Bourdieu (1977) and upon ethnomethodology, which is Heideggerian in origin (Wenger & Lave, 1991, Seely Brown & Gray, 1995) has been deployed in U.S. corporations (Wenger, 1999; Knowledge Garden, 2000).

The whole group of methods denoted "qualitative" and the theories that guide them originated in the 19th century. Max Weber, respected by virtually all sociologists, was emphatic on the point that it is delusional to believe one can describe social phenomena without, in his words describing them from a "particular point of view" in each and every case.

The more comprehensive the validity—or scope—of a term, the more it leads us away from the richness of reality, since in order to include the common elements of the largest possible number of phenomena, it must necessarily be as abstract as possible and hence *devoid* of content.... All knowledge in cultural reality...is always *knowledge from particular points of view*. When we require from the...research worker that...they should have the "point of view" for this distinction, we mean that they must understand how to relate the events of the real world...to "cultural values" and to select out those relationships which are significant for us. If the notion that those standpoints can be derived from the "facts themselves" continually recurs, it is due to the naïve self-deception of the specialist who is unaware that it is due to the evaluative ideas with which he unconsciously approaches his subject matter, that he has selected from infinity a tiny portion with the study of which he concerns himself. (Weber, 1949)

This statement stands in strong contrast to the positivism of Auguste Comte (1798-1857) and followers of this philosophy who believe that a final truth, even about social phenomena, can be reached through the methods of science, where science is understood as expressed in measurements. The term positivism derives from the belief that society will become ever more perfect as a result of advances in science, including the social sciences. Positivism, which in IS research is often equated with scientific rigor, is in fact based on a philosophical belief about what science can accomplish in the social realm (Comte, 1988; Jones, 1998). This belief also implicitly underlies the notion that technology will improve human existence. Positivism is often equated with empiricism. However, this too is problematic. Merriam-Webster's online includes the following contradictory definitions:

1: originating in or based on observation or experience <empirical data>
2: relying on experience or observation alone often without due regard to system and theory
3: capable of being verified or disproved by observation or experiment <empirical laws>

Linked to positivism are the two foundational ideas of analytic philosophy and of rationality.

Modern empiricism has been conditioned in large part by two dogmas. One is a belief in some fundamental cleavage between truths which are *analytic*, or grounded in meanings independently of matters of fact and truths which are *synthetic*, or grounded in fact. The other dogma is *reductionism*: the belief that each meaningful statement is equivalent to some logical construct upon terms which refer to immediate experience. Both dogmas, I shall argue, are ill founded. One effect of abandoning them is, as we shall see, a blurring of the supposed boundary between speculative metaphysics and natural science. Another effect is a shift toward pragmatism. (Quine, 1961)

While qualitative research can be analytic, rational and positivist (Myers ongoing), it tends to be associated within IS with more relativistic, descriptive and phenomenological approaches. This poses something of a problem in contrasting quantitative with qualitative approaches, because there is a contrast among qualitative works just as there are interesting differences amongst quantitative approaches. On the whole in this chapter, I take a position about the future of research that is descriptive and that questions the reification of many constructs upon which much quantitative research within information systems is based.

Weber's point is that social inquiry is based on interests and perspectives, and that for this reason the objectivity implied in the analytic tradition is unattainable, and the reductionism alluded to by Quine leads us farther from the "truth" about situations. In other words, social truths are approximate at best and we are best off just admitting it and stating our perspective. Weber himself could be considered an empiricist according to the first definition above, in that his work was based on extensive historical research and social observation.

C. Wright Mills (1959) develops the point about interests throughout his book *The Sociological Imagination*, a treatise on the sociopolitical structures within which sociology was conducted in his time. A contemporary view would go further to say that interests are so deeply embedded in an often contradictory context that they are only just available to be examined, even less able to control their own direction or that of society at large (Ciborra, 2000). Phenomenological theories now inform approaches to gathering and writing up accounts of situations that affect the design, development, implementation and use of information technology. These formative contexts within which IT is conceived and deployed have become more visible and more critical to the successful use of IT as a strategic tool (Ciborra, 1994).

In some recent developments, domains as seemingly "purely technical" as network infrastructures have been analyzed in terms of actor-networks, relying upon descriptive multi-dimensional accounts (Monteiro and Hanseth, 1996; Ciborra, 2000). In other words, qualitative/descriptive methods in general have encompassed a wide range of IS research issues and have shown great explanatory power. The growth of qualitative research within the discipline coincides with the shift from systems using highly structured methods such as transaction processing, to applications where groups and work practices must be understood as social processes in order for the benefit to be realized.

The ability both to explain complex themes and to chronicle detail that builds a comprehensible narrative has been the abiding strength of the qualitative approach since Weber's time. It allows consideration of a wide range of interrelationships among historical events and social structures over multiple domains, a feat that is difficult to accomplish using quantitative methods precisely because of requirement in the latter to formulate provable hypotheses and to isolate rather than aggregate variables.

However, with the advent of complex adaptive systems and chaos theory, this situation is changing. As a consequence the relationship between qualitative and quantitative research ought also to change. A more sophisticated quantitative research does not preclude a thriving vein of qualitative research, but rather opens the possibility of altogether new forms of inquiry in which it is possible these distinctions will be diminished or seen as complementary.

To explain the title metaphor: what is disconcerting about a Möbius strip or an Escher drawing is that prior expectations about topography are violated. We expect an "inside" and an "outside" to objects. Discomfort arises as we see that what contradicts common sense about matter is logically possible with lines. The difficulty of describing trends in qualitative research in IS is that temporally it is difficult to distinguish inside from outside. The determination is in some ways made locally according to a position on the line and the environment around it. The material world of embodied persons who distribute the funds for the institutions that sponsor IS careers may be at a different locality on the line from some of the research that is breaking ground in the field. This is part of our discussion. A more provocative question might be: "Are we there yet?" The answer is "Yes, but you still can't get out of the car."

QUALITATIVE RESEARCH IN THE IS CONTEXT

The problem that arises over qualitative methods in information systems is that they are discontinuous within the short history of the IS discipline and the disciplines that informed it: business economics and computer science (Lee and Liebenau, 1997). The lateral introduction of qualitative research into a research stream originating in highly structured world views that are direct inheritors of positivist philosophy (Escobar 1994) is the cause for a dilemma of legitimacy. This book and other recent efforts (*MISQ* special issue on qualitative research, IFIP WG 8.2 1997 conference on qualitative research) reflect that dilemma and address it. The dilemma is compounded by several other developments that make the timing of the qualitative debate anachronistic even as it is important. These developments include:

1) Other problems are being examined in disciplines like anthropology where the conversation has been drawn out for a longer time. These are problems like "How do we incorporate the post-modern critique and still have a discipline left?" (Fox, 1991),

2) Richly explanatory IS research drawing on post-modern theories (Ciborra, 2000; Latour, 1999a &b) exists in the same climate as the discussion about qualitative research in general *versus* quantitative research as understood in the 1970s.

3) Computing itself and other mathematically based sciences are in the process of incorporating less predictive but more comprehensive models like complexity theory and chaos theory, in some ways transcending the present debate (e.g., Gleick, 1988).

4) There are "qualitative" user interfaces like 3D screens, increasingly "hyper" text and graphics, and experiential scenarios mediated by technology that while digital, appear more and more analog, contrary to humanistic expectations (Dunn et al., 2000; Patten et al., 2000). While digital and analog don't exactly equate to quantitative and qualitative, the fact is that a similar boundary seems challenged in another area.

5) Internet users are becoming increasingly diverse and so that as transactions become virtual, outcomes are more local, particular, and embodied. For example, Peruvian expatriates order home-made cakes from local housewives to be delivered as birthday gifts to their mothers in provincial towns, and numerous other examples of homegrown uses. (World Bank, 2000).

In short, the issue under discussion is changing even as the past is being brought into the present in IS research. It was Weber who introduced the term "rational bureaucracy" into our vocabularies. The rational bureaucracy in Weber's time was one based on merit and on measurement rather than on family relations and more medieval forms of patronage. This has been a powerful notion, one that while starting as an observation became normative.

Some scholars have no doubt experienced profound surprise that the modernist project to develop technological systems for the rational bureaucracy has managed instead to illumine the organization as locally organized, socially mediated, and more effective when less than fully rational (Ciborra, 1994). The Internet itself is the ultimate embodiment of that discovery. The military structure of the command and control organization is no longer needed for coordination, but rather new coordination models have replaced it (Ciborra, 2000; Malone and Crowston, 1990). Technology seems in many ways to have outgrown its parent, rationality. That is, technology created for a rationalistic purpose using rationalistic means, allows for so much complexity, diversity and parallel processes that it now exceeds our ability to encompass it with the rational models that originally conceived it.

How do these changes affect the role of qualitative research in IS? They form the backdrop of scenarios being acted out even as we debate how to document and account for them. They are so complex that they practically demand not just qualitative research but research that can incorporate many different kinds of descriptions, measures and insights. Ethnography, held up as the model of qualitative research, is barely adequate to so complex a task, except as performed by a person of great perception and imagination, such as Latour. Latour, significantly does not isolate technical from human systems, but openly addresses the multiplicity of elements in our everyday navigation through a multi-element and often conflicted, confusing conceptual space that nevertheless is constantly active and consequential (Latour, 1996b).

In the end the discussion has gone beyond a question of methods and must instead be about the frameworks that are capable of describing a socio-technical world as it unfolds, well ahead of our ability to know what is happening. This would suggest two models that at present seem distinct, but one would guess must converge into a postmodernist version of chaos theory.

TRENDS? QUALITATIVE?

This chapter looks at the present and slightly ahead to discern what may be expected from this research approach in the future. This goal was the spirit of the request from the volume's editor, Eileen Trauth to include a chapter on new directions in qualitative research. The title phrase "trends in qualitative research" at first seems transparent, but looking closer it becomes problematic. What is qualitative research and what is a trend in it? Is there a single direction of trend or is it multiple, cyclical, spiral, or disappearing? That problem serves to illumine the situation further, because it reveals some of the assumptions held within the discipline about this research approach, about what is "current" and about the form of research most commonly contrasted with qualitative research: quantitative research.

Some people confuse quantitative research with empiricism and consider qualitative research as not empirical (Lee and Liebenau, 1997; Myers, ongoing). This is just one misconception that arises. It is based on what counts as "fact," and whether or not social, or any facts, exist independent of social context (Kuhn, 1970; Rorty, 1991). It is not clear what is understood by "qualitative research" nor exactly what it contrasts with. In practice within IS qualitative appears to stand in contrast to survey research. However, as Myers' excellent IS World page (ongoing) illuminates, qualitative research encompasses multiple research paradigms and approaches.

Another confusion may arise from the ways qualitative research was introduced into certain IS specializations. For instance, in system development, early works (Wynn, 1979; Suchman, 1987) that informed new approaches were based in ethnomethodology, which may have become equated with ethnography in general, according to Button and Dourish (1996). In other words, instances of qualitative research entered the IS arena from other disciplines.

Within their own disciplines these instances evoked certain methods, givens, theories and validity models. Imported to IS, they lost this particularity and became "qualitative research" in general. It is important to recall that the two early works cited above, being dissertations and thus sanctioned by a discipline, brought an epistemological relief to people who were trying to cope with human-centered situations without the blessing of a discipline for their first-hand observations. The 1982 IFIP WG 9.1 conference at Riva del Sole for instance (Briefs, Ciborra and Schneider, 1983) was attended by a number of union advocates who were calling for their organizations to take the measurements to disclose issues about the value of understanding work practice. It was assumed that these "measurements" needed to be "scientific" by some perhaps impossible standard, when incident logs would suffice.

The latter was a means within the grasp of the union people and the system developers who were sensitive to their needs (Wynn, 1983). This case highlights the importance of the "particular point of view." The participant's perspective was just beginning to gain legitimacy in Scandinavia, from whence it spread worldwide. But the methodology for establishing this legitimacy was not necessarily the methodology of either cognitive science nor of organizational psychology, the two legitimated "people methods" at the time. Each of these had a different slant and application, and most important were oriented towards individuals rather than groups. As well, it should not have been expected that one party ought to carry out research that appeared more in the interest of another party. Methodology was the invisible barrier to participant-based research that could solve the dilemma the union people faced.

Qualitative research became relevant in IS for the purpose of understanding users. But it developed further into studies of processes, practitioners, the organization and a reflexive view on the IS discipline itself (Orlikowski and Baroudi 1991, Lee and Liebenau 1997) on the practice of science in general (Latour 1988, Knorr-Cetina 1999) and certainly on system development practice (Fitzgerald 1997). Reflection on the practice of science and technology is important because in order for qualitative research to be deemed to have the force of "science," science itself must be disclosed to be in essence either qualitatively constituted or socially constructed (Kuhn, 1970; Berger and Luckmann, 1967), or both.

Lee and Liebenau (1997) also observe that qualitative research became a resource in information systems research because it met a specific need. It would have been difficult to introduce a change of paradigm in the absence of anomalies and unsolvable yet evident problems (Kuhn, 1970) not addressed by the paradigm. They refer to the notion of breakdown, through the metaphor of a bridge whose material properties become observable when the bridge is stressed to the point of collapsing. The bridge they refer to is the quantitative methods in use in the IS field.

But the breakdown metaphor applies ontologically as well. A breakdown is one way to become explicitly aware of the properties and capabilities of something, for instance, a tool. One of Heidegger's key notions is the concept of ready-to-hand as opposed to present-at-hand. Tools that are "ready-to-hand" are within reach and applied without reflection. If the tool is not ready-to-hand, then a reflection arises, as to the nature of the tool that is needed, and the properties of a tool that would work. This is the condition of "present-at-hand." We find ourselves now in a situation of "present-at-hand" with regard to what tools are appropriate for understanding the creation and use of information technology.

Heidegger's view of "method" and its relationship to technology is also pertinent here. Method as such is traced to Descartes, within the context of religious attainment, that is, the idea that salvation can be attained through a set of techniques. Philosophically this is seen as setting a stage for enlightenment views that predate our own search for method and the origin of the modern concept of technology. Heidegger's analysis, whether taken literally or metaphorically, is illuminating to our purpose here, since we are considering types of research that apply to the understanding of developments in information technology. The following quote is taken from Zimmerman (1990) and includes quotes from Heidegger.

> No entity can resist or stand on its own in the face of the self-reflective, anthropocentric power drive of human reason. Only being as such can never become objectified. Hence, Heidegger called for a new kind of thinking, a thinking which "let entities be" and which recalled being as such. Instead of existing in the "openness" of being so that entities could manifest themselves appropriately, technological humanity fills up that openness with methodological projects which compel entities to manifest themselves in a one-dimensional way: "What presences the entity does not hold sway but rather

assault rules.... Representing is making-stand-over-against,
an objectifying that goes forward and masters." ...For
Heidegger, modern science was the best example of this
entrapping, ensnaring way of disclosing things.

This entrapping, ensnaring way of disclosing things is altogether
consistent with the control motives of social science research and of the
positivist agenda in general. While we can't argue with the entire
premise and thrust of our research traditions nor with the development
context that funds them, we can notice when they have gone too far.
That is the point at which the bridge no longer bears the weight of the
material it must carry. The production of information technology has
undoubtedly created a crisis in terms of the speed at which ideas about
its interaction with society must adapt. The juncture practically compels another view as the basis for our reasoning about social and
technological change. That view must be indeed a "view" that is an
inspection at the local level, before the generalizations and abstractions
for large-scale quantifications can occur. This is not a quick process but
requires building up of insights from numerous studies. These studies
will not be "comparable"; they will be descriptive and insightful. Even
within these guidelines they will also of necessity be questioning of
definitions that no longer apply.

For instance, we need to examine our assumption that "technology" and "the social" are separate entities (Bloomfield and Vurdubakis,
1994). Everything is encompassed in "the social"; and boundaries
created to draw distinctions are problematic in any given instance.
Similarly the tools for studying and creating those boundaries in
intellectual endeavors are problematic. The entire problem arises
historically from a specialized role created for quantitative research
within the "modernist project" (Escobar, 1994). Because of all these
issues, it is important to define qualitative research in terms of its
historical role in social science as a background for its role in information systems management.

We should also examine the idea of a trend as it could apply to this
development. As qualitative research comes into the IS research
practice, it both pulls forward the past (of practices that are already
mature in other disciplines) and collapses into the future (of what is
developing elsewhere while the point is being debated in IS). There is
a "warp" in "progress" whereby advanced quantitative methods such as
chaos and complexity theories have moved so far beyond the simple

statistical methods of past social science that they make "quantitative" different from what it was, and they also introduce so much uncertainty into prediction that qualitative research can play a strong role in explaining processes that quantifications do not resolve.

CULTURAL AND INTELLECTUAL HYBRIDS AND POSTMODERNISM

Escobar (1994) speaks of "hybrid cultures" in the anthropology of modernity. That is, that rather than supposing the existence of dichotomous traditional versus modern cultures, in any society today we can find elements of many historical patterns intersecting within and across groups. This undoubtedly applies to the research culture of IS and to the practitioner culture as well. Indeed, Escobar's work is an example of one trend in qualitative research, which is reflexive and critical. "Hybrid" generally describes the condition known as postmodernism whereby elements from multiple historical periods are incorporated into a style that nonetheless appears as "designed."

While many works show IS research to be in a postmodern condition (Latour, 1999; Ciborra, 2000; Hanseth and Monteiro, 2000), at the same time, "traditional" IS still maintains a focus on quantitative research as the dominant paradigm. This is not surprising. The political structures and reward systems are set up to support this approach despite the widespread practice of qualitative research and innovative theory (Mills, 1956). The senior members of the communities granting recognition understand it (Rabinow, 1999) and the authority of the IS discipline derives from two quantitative disciplines, economics and computer science (Lee and Liebenau, 1997). Deans in the business schools that most commonly house information system departments — unless they are in schools of computing or library science, also highly structured practices — rarely have graduate degrees in information systems. Although they may be aware of the "social" or "human" side of things, in the end their decisions tend to be guided by conservative criteria, the more so because the IS discipline is relatively new and needs to be differentiated from other business school curricula (King and Applegate, 1997).

The history of business schools and their *raisons d'être* also lend themselves to greater conservatism. While the departments of sociology became radicalized in the 1960s, schools of business pursued organizational research from the perspective of advising management. This is perfectly consistent with the purpose of the schools. It also served to perpetuate a positivist research agenda in the sense of having the goal of controlling human destiny with scientific discovery and heading toward an ever more rational and therefore positively evolving world to meet their needs and plans. Methods come packaged with this approach. So in this agenda, the appearance of science is important to the achievement of the end. Both the idea of being able to control outcomes and the representational rhetoric of numbers form the basis for the credibility and utility attributed to the results of quantitative research. Escobar (1994) outlines a large set of disciplines engendered by the post World War II version of positivism on the part of the Allied powers and especially the United States. This agenda was to exercise international political control through "development economics."

Basing itself on Comte's premise of the perfectibility of society through science, development economics encompassed an array of social science disciplines and fostered research grants towards the end of eradicating poverty, illiteracy and disease. Noble as these goals are, their coupling to Cold War funding and government action constrained the entire effort into what Escobar (1994) calls a "development ideology." How does this affect us today? This same philosophy of social research has been applied to the organization. Mills emphasizes that the tie-in between funding rationales and quantitative positivist research cannot be overemphasized. Since technology in most cases is an implementation of this agenda for perfectibility of society through "progress," most technological interventions are formed within this context, and they certainly are funded that way.

However, the approaches that informed qualitative research in information systems have existed in parallel to positivist methods during the entire history of the social sciences. The authority of Weber never receded in the social sciences either for the positivist or for the more descriptive schools of thought. Marxist theory was positivist in its hopes of a progressive society that through logic would provide equal goods to all. But in its implementation as a political philosophy, the goals of Marxist study inevitably came into conflict with institutionally funded social science research that tended to serve establishment ideologies. The critical theorists incorporated an interpretivist epistemology into an activist emancipatory agenda. This theme was one of the first to take foothold in information systems. Linked with the social construction of knowledge and ethnomethodology, qualitative research entered the information systems field in a body of methods that addressed a wide range of concerns about the impact of technology on society and on the workplace. There were three main themes in the introduction of qualitative research into IS.

1) One drive in adopting qualitative research was to understand issues that were not accounted for in the constructs of established research. The field was in a rapid state of growth and many unprecedented issues were appearing. There was no vocabulary or observational basis for jumping to quantification. In other words, proper empiricism wasn't possible under the circumstances since there was no known relationship between the constructs and their indices or measurable phenomena.

2) Another goal was an intellectual search that questioned the model of truth that was operational. Information systems are introduced into a social and organizational environment. They are created by this environment and in turn affect it. This process is not easy to reduce to a formula, though attempts were made. Thus the relationship between IT and the organization required a new understanding of how organizations work. In fact, a new understanding of organizations emerged just from the breakdowns of technology that resulted from the mismatch of plans and outcomes.

3) A third goal was advocacy. It was not just union representatives who had concerns about technology's impact on work life. It became a managerial and an executive question as well. Impact on the workforce had many forms and many angles of interest. In the process, the role of work groups and work teams in knowledge creation came to light.

The major impulse to introduce qualitative methods came from those espousing Habermasian goals of emancipation (Hirschheim and Klein, 1994), or those who sought to avoid the degradation of social equity that might arise from technology. But these were soon accompanied by many others who saw problems arising from both the introduction of technology and the means of documenting it.

CHANGE OF TRAJECTORY

This shift in method is not unique in the study of human institutions. Even without explicit attention to political objectives, it is possible to trace a Hegelian dialectic in the focus of the social sciences themselves, summed up most poetically by Latour (1999b).

It is not exactly true that social sciences have always alternated between actor and system, or agency and structure. It might be more productive to say that they have alternated between two types of equally powerful dissatisfactions: when social scientists concentrate on what should be called the micro level, that is face-to-face interactions, local sites, they quickly realize that many of the elements necessary to make sense of the situation are already in place or are coming from far away; hence, this urge to look for something else, some other level, and to concentrate on what is not directly visible in the situation but has made the situation what it is. This is why so much work has been dedicated to notions such as society, norms, value, culture, structure, social context, all terms that aim at designating what gives shape to micro interaction. But then, once this new level has been reached, a second type of dissatisfaction begins. Social scientists now feel that something is missing, that the abstraction of terms like structure and culture, norms and values, seems too great, and that one needs to reconnect, through an opposite move, back to flesh and blood local situations from which they had started. (Latour, 1999b, p. 17)

This meta-view of the social sciences has been noted before (e.g., Mills, 1959, and many more) but probably never explained as above. That is because in the past macro and micro, which in the minds of many are the topical versions of quantitative and qualitative, in essence, have been seen as layered, if connected at all. Latour makes the case that not only are they telescoping views of embedded phenomena, but also that the activity at the local level both forms and reflects the structure at the abstract or larger "network" level. Actor network theory (ANT) as Latour describes it (he grudges calling it a theory, but we can't go into that here) involves anything that has a name in our experience and the relationships among those named distinctions, as well as their perceived ordering or consequentiality to one another. Actors can be objects, or even ideas; the network is the movements or "circulations" between them.

> It seems to me that ANT is simply a way of paying attention to these two dissatisfactions [named in the previous quote], not again to overcome them or to solve the problem, but to follow them elsewhere and to try to explore the very conditions that make these two opposite disappointments possible. By topicalizing the social sciences' own controversies, ANT might have hit on one of the very phenomena of the social order: maybe the social possesses the bizarre property of not being of agency and structure at all, but rather of being a circulating entity. The double dissatisfaction that has triggered so much of the conceptual agitation of the social sciences in the past would thus be an artefact: the result of trying to picture a trajectory, a movement, by using oppositions between two notions, micro and macro, individual and structure, which have nothing to do with it.... ANT refers to something entirely different [from what we have conceived as the polarities above] which is the summing up of interactions through various kinds devices, inscriptions, forms and formulae, into a very local, very practical, very tiny locus (Latour, 1999b, p. 17)

Here is where we make a metaphoric leap. In describing some of the virtues of qualitative research, we have named the importance of understanding the local, in its detail and particularity, as a starting point for any possible generalization. Latour has referred above to "a very local, very practical, very tiny locus."

The "local" or "local organization" as it happens, is also an organizing principle in chaos theory. That is, given a set of identified vectors in a system, the complexity of that system makes it impossible to predict their interactions. This is more the case, the more variables are considered. Given that there are very many variables involved in global technology development, some of which are named in the world systems field of sociology cited below (Hopkins and Wallerstein, 1996), it becomes increasingly difficult to predict how they will interact. Indeed, vectors in chaos theory are anticipated to interact separately amongst each other and to organize locally to affect the balance of the entire system. World systems theorists (Hopkins, 1996) have used this as a premise for anticipating a change in the world system, based upon interactions among the following set of vectors:

- The interstate system
- The structure of world production
- The structure of the world labor force
- The patterns of world human welfare
- The social cohesion of the states
- The structures of knowledge

"None of the six vectors has, or in our view, could have, developed in isolation from the others. The vectors are not at all to be thought of as loci of autonomous forces. They form, rather, the minimum array of interrelational facets of a single, imperfect, organic whole, each vector quite dependent on the others. Any shock or blockage, or transformation within any one of them or among them, affects all the others, and usually soon, visibly and consequently" (Hopkins and Wallerstein, 1996, p. 2)

Although the list of vectors doesn't explicitly include technology, to anyone involved in this field, it would seem to jump out as a factor. Technology has become a major means of production, and through informatization (Zuboff, 1996) its locus of control is also changed. The quote is included here to illustrate how chaos theory can be constructed for the purpose of studying social phenomena. Thus, ironically for our present debate about the virtues of the qualitative, current developments in the physical, mathematical and the social sciences show an unexpected convergence or accommodation between what have been understood as quantitative and qualitative methods. Positivism remains a tacit system of belief, but science no longer needs it or even espouses it openly.

Chaos theories have entered the management arena in the form of popular management books (e.g., Brown and Eisenhardt, 1998; Sanders, 1998; Wheatley, 1992). Wheatley's (1992) book directed at a corporate audience is now in its fifth printing. While the implication of popular books is that corporations can continue at least to steer their futures by embracing change and chaos (still positivist), the clear message is that traditional methods of planning and the hope of control are obsolete. This message is easy to grasp intuitively given the condition of markets and global changes most large corporations have direct experience of. Whether or not readers fully grasp the implications of chaos for their planning purposes is not known. Ciborra's newest book as of this writing (Ciborra, 2000) is even more direct in its very title *From Control to Drift: The Dynamics of Corporate Information Infrastructures*. Its amazon.com blurb introduces this notion more directly: "Far from being a linear process, the use of the information infrastructure is in fact open-ended, in many cases out of control. Current management models and consulting advice do not seem to be able to cope with such a business landscape." This is not to say that there is nothing corporations or institutions can do to succeed and to manage chaos and complexity. But it does suggest that a very different premise and understanding about how complex systems work is required. This is not necessarily consistent with reductionist approaches of the past.

Mathematical models used for chaos and complexity studies (Kaufmann, 1995; Gleick, 1998) span many disciplines. It is the complexity of interactions between the constituents of a system that generate chaos, not necessarily their size or number of variables. However, when there are many variables and when initial conditions are difficult to establish, the conditions for chaos are present. It can be presumed in technology change that these original conditions will almost always be changed by small increments and that therefore end results are highly unpredictable. Here is where explanatory models based in social interaction and actor networks can offer something that eludes quantification under the circumstances, which is a way of perceiving the dynamic at the local level and therefore making good guesses about how these might play out on a large scale.

Local organization is the key property of chaotic systems. They are so complex that their vectors or variables interact among each other in locally organized unpredictable ways. These small-scale local organizations have large-scale systemic effects because of the amplification provided by the shifts involved in the interactions and the size of the systems. At the other end, the ethnomethodological and actor-network end (if indeed they are on a spectrum, now cast into doubt), we also have a bigger "structure" which is the effect of local interactions in the construction or exchange of meanings among actors, otherwise known as the network. The question left us then is "Are these the same kinds of local effects?" Does the rigor of mathematics ultimately lead to similar principles as the reflections of phenomenologists? Or is this too much to hope for? Regardless, contradictions will continue because they appear built into the social fabric of science, into political interests and into the search for control and the accommodation of change.

SOME POSSIBLE RESEARCH TRENDS

One trend is simply to refine the fit between methods, understanding and practice: using the understandings of qualitative research in the design of technology and of the institution. Another trend is also already ongoing: the incorporation of qualitative methods into the discourse of modernity. Ethnography is now routinely applied not just to tribal, village and ghetto environments, but also to mainstream problems of complex organizations. In this sense the "otherness" of early anthropology has dissolved into a theme that implies that the setting of modernity is just another human setting rather than being a privileged one. Thus the computer systems development environment, nothing if not an icon of modernity, is amenable to exactly the same methods of understanding as the tribal camp. This minimizes rather than maximizes the distance between industrial and pre-industrial society, at the same time that it attempts to solve problems that specifically arise in the industrial environment.

Legitimation is certainly an issue in the trends for qualitative research. The increasing availability of journals and special issues that encourage the submission of papers based on qualitative research is one such legitimation. Another is the corporate adoption both of user-centered design concepts and of knowledge management efforts. Concurrently the markets for technical products have long since passed the point where sheer computing power was a sufficient differentiator. In the emerging applications more and more sensitivity to user contexts is needed in order to create markets for products that are addressed, for instance to the consumer, but also to more subtle organizational practices.

Intel Corporation, among others, has legitimated user-oriented design by creating a job classification of "human factors engineer." While the extension of the term engineer to a wider category of activity may seem to impose a positivist framework on a broader realm of inquiry, at the same time it also establishes the credibility of new areas of work within the respected and known classification of "engineering." The broadening of the umbrella of engineering to human factors covers far more than user interface design. It covers every user-oriented form of research that informs products and systems. This legitimation into a job classification is a powerful enabler for more extensive application of a broad range of methods to studies both on products and on organizational solutions.

Trends in qualitative research may actually link qualitative methods back to mathematical models, now that the modelers have softened on the premise of absolute predictability and control. Modeling now has an acknowledged indeterminate element that can in the right application be philosophically sympathetic to the motivation behind qualitative research. That is, the search for the dynamics of local organization may have to be made observationally; and in human systems they may need to be participatory, since it is possible they are meaning-based.

Emerging forms of complex modeling conversely result in systems so flexible in their use that the effect can be considered qualitative. In these areas there is a cross-disciplinary perception of the potential of applying methods of computer science and information science to problems so complex that there is no way other than sheer intuition and experience to "analyze" them. Yet the calculation capabilities have become so great that it makes sense to produce information clusters that change constantly and whose applications are entirely "interpreted" at the user end.

There are many such programs, especially in the new knowledge and visualization research areas, where many thousands of documents can be classified and then represented in a visual topography that locates like items in the same "region." For example, there are indexing programs that perform semantic analysis using algorithms to cluster like content from thousands of documents. Access to the content however is navigated through the user's sense of how things are similar. In other words, a very analytical approach makes it possible to handle large volumes of information in such a way that intuitive access is possible. In fact, the underlying processing power of the programs eliminate the need for the user to retain a set of search keys that would lead them into a topic. In addition, this information can be constantly updated so that new topographies evolve daily. The point is not to advertise the programs, but to illustrate that analytical tools based in mathematics and logic can pave the way to what we might call "qualitative" uses, if we were still employing that language.

At the other end of the spectrum, we have theories totally removed from the preoccupation of addressing a factual world and that address entirely the construction of terms and the combination of elements involved in that construction. As an esthetic and an interpretive framework, actor network theory is a very strong candidate for a potent trend in qualitative research. A difficulty with contemporary theories and one expressed for actor-networks is that of trying to force-fit them into prior views of what we thought a theory ought to accomplish.

> The notion of a network is itself a form—or perhaps a family of forms—of spatiality: that it imposes strong restrictions on the conditions of topological possibility. And…accordingly it tends to limit and homogenize the character of links, the character of invariant connection, the character of possible relations, and so the character of possible entities…Actor-network is, has been, a semiotic machine for waging war on essential differences. It has insisted on the performative character of relations and the objects constituted in those relations…(Law and Hassard, 1999, p.7)

The positivist, or as Law (1999) expresses, the "managerialist" (control-oriented) tendency remains, even when the theory has declared these agendas unrealizable. "What do we do with this?" remains as the essential question. What is appealing about actor-networks is the lack of distinction between social and technical, or between any other established pairs of meanings. They are all encompassed within the semiotic, without arguing about the physical substantiality of any element. Yet the supposition of a "network" already establishes a kind of boundary around the possible topics.

CONCLUSION

So, as qualitative research gains ground in contrast to quantitative research within the IS discipline, quantitative research itself is drastically changing. Concurrently there is a shift in the broader thinking and understanding of corporations and specifically IT groups, to premises much closer to those described in the postmodernist or phenomenological worldviews. This is because they face highly complex and unpredictable market situations and recognize that methodological conservatism is a potential threat to their ability to thrive in these new circumstances.

Our practical inability to structure phenomena in conventional ways because of the fast and self-accelerating nature of socio-technical changes compels an openness to approaches that have been available for some time. Neural networks, communities of practice, peer-to-peer computing relationships, aggregating network infrastructures (Ciborra, 2000) are themes in corporate IT thinking. What to do with these ideas is another question, but the fact is that their discontinuous premises are accepted as possible solutions in a way that would have seemed impossible in the 1970s.

REFERENCES

Berger, P. and Luckmann, T. (1967). *The social construction of reality: a treatise in the sociology of knowledge*, London: Penguin Publishers.

Bloomfield, Brian P. and Vurdubakis, Theo (1994). Boundary Disputes: Negotiating the Boundary between the Technical and the Social in the Development of IT Systems, *Information Technology & People*, 7(1), 9-24.

Bourdieu, Pierre (1977). *Outline of a Theory of Practice*. Cambridge Studies in Social and Cultural Anthropology, 16.

Briefs, Ulrich, Claudio Ciborra & Leslie Schneider, Eds. (1983) *Systems Design for, with and by the Users*. Amsterdam: North-Holland Publishing Company.

Brown, Shona L. & Kathleen M. Eisenhardt (1998). *Competing on the Edge: Strategy as Structured Chaos*. Cambridge, MA: Harvard Business School Press.

Button, Graham and Paul Dourish (1996). Technomethodology: Paradoxes and possibilities. *Proceedings of CHI 96*, ACM, New York.

Ciborra, Claudio U. (2000) *From Control to Drift: The Dynamics of Corporate Information Infrastructures*. Oxford UK: Oxford University Press.

Ciborra, Claudio (1994). The grassroots of IT and strategy. In Ciborra, Claudio and Tawfik Jelassi, Eds. *Strategic Information Systems: A European Perspective*. Chichester UK: John Wiley & Sons.

Ciborra, Claudio and Tawfik Jelassi (Eds.) (1994). *Strategic Information Systems: A European Perspective*. Chichester UK: John Wiley & Sons.

Comte, Auguste (1988). *An Introduction to Positive Philosophy*. Hackett Publishing Co.

Escobar, Arturo (1994). *Encountering Development*. Princeton University Press.

Fox, Richard G. (1991) *Recapturing Anthropology*. Santa Fe, NM: School of American Research Press.

Gleick, James (1988). *Chaos: Making a New Science*. New York: Penguin Books.

Grint, Keith; Case, Peter; and Willcocks, Leslie (1996). "Business Process Reengineering Reappraised: The Politics and Technology of Forgetting" in Orlikowski, W.J.; Walsham, G; Jones, M.R. and DeGross, J.I. (Eds) *Information Technology and Changes in Organizational Work,* Chapman & Hall, London.

Habermas, J. (1984). *The Theory of Communicative Action: Reason and the Rationalization of Society,* Vol. 1, T. McCarthy (tr.), Boston, MA: Beacon Press.

Hanseth, Ole & Eric Monteiro (2000). Understanding Information Infrastructure. Available on the web at http://www.ifi.uio.no/~oleha/ Publications/

Hirschheim, R. and Klein, H. (1994) "Realizing Emancipatory Principles in Information Systems Development: The Case for ETHICS," *MIS Quarterly*, 18(1), 83-109.

Hopkins, Terence K. and Immanuel Wallerstein (1996). *The Age of Transition: Trajectory of the World-System 1945-2025*. Zed Books, London & New York.

Jones, H.S. ed. (1998). *Comte, Early Political Writings. Cambridge Texts in the History of Thought.* Cambridge University Press, Cambridge, UK.

Kauffman, Stuart (1995). *At Home in the Universe: The Search for the Laws of Self-Organization and Complexity*. Oxford, New York: Oxford University Press.

Knorr-Cetina, Karin (1999). *Epistemic Cultures: How the Sciences Make Knowledge*. Cambridge, MA: Harvard University Press.

Knowledge Garden (2000). http://www.co-i-l.com/coil/knowledge-garden/cop/companies.shtml.

Kuhn, Thomas S. (1970). *The Structure of Scientific Revolutions (2nd edn)* University of Chicago Press, Chicago.

Latour, Bruno (1999a). *We Have Never Been Modern*. Cambridge, MA: Harvard University Press.

Latour, Bruno (1999b). On recalling ANT. In Law, J. and Hassard, J. (Eds.). *Actor Network Theory and After*. Oxford UK: Blackwell

Publishers.

Latour, Bruno (1988). *Science in Action: How to Follow Scientists and Engineers Through Society.* Cambridge MA: Harvard University Press

Latour, Bruno (1996a). *ARAMIS or The Love of Technology,* Cambridge MA: Harvard University Press.

Latour, Bruno (1996b). Social Theory and the Study of Computerized Work Sites in Orlikowski, Wanda J., Walsham, Geoff, Jones, Matthew R. and DeGross, Janice I. (Eds) *Information Technology and Changes in Organizational Work.* London: Chapman & Hall.

Law, John & John Hassard (1999). *Actor Network Theory and After.* Oxford, UK: Blackwell Publishers.

Lee, Allen S. and Jonathan Liebenau (1997). Information Systems and Qualitative Research. In Lee, Allen S, Jonathan Liebenau and Jan de Gross, Eds. *Information Systems and Qualitative Research.* London: Chapman & Hall.

Lee, J., Dunn, B. S.Ren, V.Su, and H.Ishii (2000). GeoSCAPE: 3D Visualization of On-Site Archaeological Excavation Using a Vectorizing Tape Measure, in *Conference Abstracts and Applications of SIGGRAPH '00* (New Orleans, Louisiana, July 23-28, 2000), ACM Press, 206.

Malone, T. W. and Crowston, K. (1990). What is coordination theory and how can it help design cooperative work systems? In D. Tatar (Ed.), *Proceedings of the Third Conference on Computer-supported Cooperative Work,* Los Angeles, CA: ACM Press, 357-370.

Merriam-Webster's Online Dictionary http://www.m-w.com/cgi-bin/dictionary.

Mills, C. Wright (1959). *The Sociological Imagination.* The Free Press of Glencoe IL.

Monteiro, Eric and Hanseth, Ole (1996). "Social shaping of Information Infrastructure: On Being Specific about the Technology" in Orlikowski, Wanda J., Walsham, Geoff, Jones, Matthew R. and DeGross, Janice I. (Eds), *Information Technology and Changes in Organizational Work,* London: Chapman & Hall.

Myers, Michael (ongoing). ISWorld page on Qualitative Research http://www.misq.org/misqd961/isworld/index.html#Qualitative Research Methods.

Orlikowski, W.J. and Baroudi, J.J. (1991). Studying Information Technology in Organizations: Research Approaches and Assumptions.

Information Systems Research, 2(1), 1-28.

Patten, J. and H. Ishii (2000). A Comparison of Spatial Organization Strategies in Graphical and Tangible User Interfaces, in *Proceedings of Designing Augmented Reality Environments (DARE '00)*, (Elsinore, Denmark, April 12-14, 2000), 41-50.

Quine, W.V.O. (1961). *From a Logical Point of View*. Second revised edition Harvard University Press, Cambridge, MA.

Rorty, Richard (1991). *Essays on Heidegger and Others*. Cambridge, UK: Cambridge University Press.

Sanders, Irene T. (1998). *Strategic Thinking and the New Science: Planning in the Midst of Chaos, Complexity, and Change*. New York: The Free Press.

Suchman, Lucy (1987). *Plans and Situated Actions: The Problem of Human-Machine Communication*. Cambridge University Press.

Waldrop, M. Mitchell (1992). *Complexity: The Emerging Science at the Edge of Order and Chaos*. Touchstone Books, New York.

Walsham, G. (1997). Actor-Network Theory and IS Research: Current Status and Future Prospects in Lee, A.S; Liebenau, J.; and DeGross, J.I. *Information Systems and Qualitative Research*. London: Chapman & Hall.

Wastell, David G. (1996). The fetish of technique: Methodology as a social defence. *Information Systems Journal,* 6(1), 25-40.

Wenger, Etienne and Jean Lave (1991) *Situated Learning*. Cambridge, UK: Cambridge University Press.

Wenger, Etienne (1999). Learning, Meaning and Identity. Cambridge, UK: Cambridge University Press.

Wheatley, Margaret (1992). *Leadership and the New Science: Discovering Order in a Chaotic World.*

Wilson, Edward O. (1998). Back from Chaos. *The Atlantic Monthly,* March. Available online at http://www.theatlantic.com/issues/98mar/eowilson.htm.

World Bank (2000). http://www.infodev.org/stories.html.

Wynn, Eleanor Herasimchuk (1979). Office Conversation as an Information Medium. Unpublished PhD dissertation. Department of Anthropology, University of California, Berkeley. Available through UMI/Bell & Howell.

Wynn, Eleanor (1983). The user as a representation issue in the U.S. In Briefs, Ulrich, Claudio Ciborra and Leslie Schneider (Eds.), *Systems Design For, With and By the Users*. Amsterdam: North-Holland

Publishing Company.

Zimmerman, Paul (1990). *Heidegger's Confrontation with Modernity, Technology, Politics and Art.* Bloomington: Indiana University Press.

Zuboff, Shoshana (1996). The emperor's new information economy. In Wanda Orlikowski, Geoff Walsham. Matthew Jones and Janice deGross, Eds. *Information Technology and Changes in Organizational Work.* Proceedings of the IFIP WG 8.2 working conference on information technology and changes in organizational work, Cambridge UK, December 1995. London: Chapman & Hall.

Issues
for the
IS
Researcher

Chapter III

Action Research: Helping Organizations to Change

Enid Mumford
Manchester Business School, UK

INTRODUCTION

This chapter has been written to provide some guidance to researchers who wish to be actively associated with assisting change taking place in the organizations that they are studying. This means that they will not be detached observers interested only in observation and documentation, but will take on a role as a facilitator of the change process. This can be done through advice to management, in which case the researcher is operating in a manner similar to a consultant, but it can also be done more democratically. My own approach, which has been developed over many years, is to assist the future users of any new system, together with those who will be affected by its use, to play an important role in the systems design process. This role can take different forms but usually incorporates the design of the organizational structures that will surround new technology. Many groups have now become designers of their own systems ranging from clerks, salespeople and technologists to senior managers.

ACTION RESEARCH—WHAT IS IT AND WHY DO IT?

Action research is concerned with change. Its intention is to change situations in ways that are seen as better, either by the researcher or by groups in the research situation, and to draw some theoretical conclusions from this process. This means that it can have a political element associated with it related to the question: "What is better for whom and who makes this judgment?" Action research requires the researcher to obtain an accurate and comprehensive understanding of the situation being addressed before taking any action directed at solving identified problems. Today, this may even be its principal objective. Giddens tells us that many problems are now so complex that while they can be clarified and better understood, they are too difficult to be solved (Giddens, 1991).

Action research requires a more intimate and longer term relationship with the research situation than is usual with research methods such as attitude surveys. It differs from consultancy in that one of its major aims is a contribution to both practical and theoretical knowledge. Also it may have many different groups as its principal clients whereas the client of the professional consultant is usually management. It is often defined as a method associated with qualitative research but it does not necessarily exclude statistical analysis. The wise action researcher who wishes to demonstrate that beneficial change has been an important consequence of remedial action will ensure that careful statistical records, which may include attitude surveys, are made of the pre-change situation. This enables comparative measures to be made after change has been implemented.

A method close to action research is participant observation in which the researcher tries to obtain an in-depth knowledge of a group or situation through becoming a part of it. The researcher can either be a concealed observer, a role that is not dissimilar to that of the spy, or she can be quite open about what she is doing, asking the group if she can work with them so as to better understand the nature of their work and its problems. Professor Andrew Pettigrew used this method when he collected data from programmers and systems analysts for his seminal study of organizational politics (Pettigrew, 1973; Mumford and Pettigrew, 1975). My own research has usually included an element of participant observation although very early research did not include computer systems. Few had arrived on the business scene in the 1970s.

An important form of action research is that carried out by the early Tavistock pioneers of the socio-technical approach. In their study of coal mines in the North East of England they spent a long time observing the organization of work underground and came up with recommendations which involved more flexible methods of working, including multi-skilling and giving responsibility for the organization of group tasks to the miners. In this research the Tavistock role was that of expert advisers and they did not involve the miners in the analysis of problems and the identification of solutions. At a later stage they became wedded to a more democratic approach in which the workers themselves developed the solutions, although they were usually guided to a socio-technical form of organization by the experts (Emery, 1993).

My own research on information systems has always tried to follow the approach of the Tavistock social scientists by helping groups to analyse their own problems and develop their own solutions. This meant that I had to maintain a number of action research roles at the same time. I had first to remember that one of my primary aims was making a contribution to knowledge and so I had to gain an in-depth, comprehensive understanding of the structure, relationships and dynamics of the research situation which could later be published to help other researchers. I had also to act as a facilitator helping the design group to analyse its own problems and arrive at solutions to remove or alleviate these. As all these groups were concerned with the introduction of new technology, this was an important factor that had to be taken account of in the diagnosis of problems and development of solutions.

The facilitator role is not an easy one and will be discussed in more detail in the case studies. It involves helping the group to choose or develop an appropriate methodology for the work ahead, to keep them interested and motivated in the design task, to help them to resolve any intergroup conflicts, and to make sure any important design factors are not forgotten or overlooked. The facilitator must in no circumstances make decisions for the design group or persuade them that certain things should be done or not done. Her role is to help the group to systematically analyse their own needs and problems and arrive at technical and organizational solutions that solve the problems and meet the needs (Mumford, 1995, 1996).

Professor Brian Gaines in his keynote speech at INTERACT in Edinburgh in August, 1999, said that "if design issues relating to technology and societies are to be understood, we need a much greater overt understanding of the operation of our societies, their economies, politics and cultures, and how these evolve under the influence of environment factors including the development of information technologies." It is this that action research can provide. It is clear that much of industry is undergoing dramatic changes in structure, functions and organisation at the present time. These changes, and their consequences are little understood and hard to manage. For example, companies are moving away from hierarchies to networks (Castells, 1996) and from centralised to decentralised structures in which parts of a company are run as semi-autonomous units. But many are pulled in conflicting directions. Managing complexity requires flexibility and diversity while profit generation requires efficiency and control. These two sets of needs are difficult to combine. Also networks and democracy often run counter to the ideology of capitalism where the objective of industry is profit for shareholders.

A report by the British Department of Trade and Industry suggests two alternatives for the enterprises of the future (DTI,1999). One is that of 'wired world' organizations. These are small companies in which networks of self-employed individuals come together via the Internet to work on common projects based on temporary work contracts. The other is the 'built to last' company. These are stable, relatively large firms not very different than the successful companies of today but operating in global markets. Their principal aim is to prosper through the collection and use of knowledge. Both of these types of organizations will be accompanied by a growth of service industries which may provide the bulk of work for a majority of people. All will require 'knowledge workers' who will be self-employed, skilled at selling themselves and at protecting their own knowledge (Handy, 1994).

Giddens suggests that these organizational changes will radically alter our work experiences, our social lives and our ideas of self identity (Giddens, 1991). Yet, at present, we can only speculate what will happen and how we will react to this dramatic change. The future is a high-risk situation (Beck, 1992). Giddens points out that this makes notions of trust very important. At present, we trust what we know, see and experience. In the future we may be asked to trust people and events that are far away and have unknown consequences. These distant happenings can have an immediate effect on our present and future lives, yet we will be unable to predict what they will be. This means that a great deal of our present knowledge will be irrelevant to our future problems.

Leadbetter points out that, in the future, there will be a great need for social responsibility and a business world that is not driven solely by the demands of shareholders and financial markets (Leadbetter, 2000). He argues that the bright, young, intelligent employees of the future will want to work for companies with good social reputations. Public trust will also be of great importance to the development of knowledge. Innovation carries risk and the more radical the innovation, the greater the perceived risk. Companies must be able to create this trust by being seen to act responsibly, yet their ethical dilemmas are likely to increase. The recent experiences of Shell and Monsanto, the first with the disposal of redundant oil rigs, the second with genetically modified crops, illustrate how public disapproval can harm profitability. The public wants to be able to trust major companies to behave ethically.

An important response to the uncertainty of the future must be more research, especially interdisciplinary research. Action research with its long-term perspective and ability both to gain an understanding of complex problems and perhaps make a solution to addressing some of these, will be of great importance. Technological developments have always been a generator of risk and uncertainty; their ability to do this in the future will greatly increase. The information systems community needs to be a leader in undertaking research.

PARTICIPANT OBSERVATION AS A PRELIMINARY TO ACTION RESEARCH

Although the emphasis of this chapter has to be on action research which is related to information technology, lessons can be learned from other methods, in fact it is important to do this. My first research projects took place when I was a member of the Department of Social Science at Liverpool University. At that time I was very influenced by the work of anthropologists and believed that a true understanding of what took place in industry required a close association with people and events. I believed that, like the anthropologist, I should live and work with the groups I was studying so that I could learn about their emotional and organizational problems in the work situation (Whyte and Hamilton, 1965).

Entering the Research Situation

My first opportunity to do this was on the Liverpool docks. The dockers, or stevedores as they are sometimes called, were constantly on strike and the National Dock Labour Board who employed them, together with the stevedoring companies who hired them from the Board to load and unload ships, wanted to know why. The Liverpool dock estate was situated at the mouth of the River Mersey and had three kinds of marine traffic. The North end handled very large ships, the South smaller ones, carrying cargo such as grain and bananas. There was another smaller dock estate across the river in Birkenhead which also handled smaller boats (Simey, 1954). I decided that I must get a job on the dock estate so that I could talk to the dockers.

My first problem was to find a female role that would be seen as legitimate and bring me into contact with the dockers. There were two roles available. One was boat scrubber. These were ladies, usually elderly, who scrubbed out the interior of the boats while they were in dock. But this group would have little to do with the dockers. The other group was canteen assistants. There were many dock canteens on the dock estate as the men had to eat between shifts. These were owned by catering companies often supporting good causes. For example, a number were owned and run by the Womens' Temperance Union. I decided to work in three of these canteens—one at the South end of the docks, one at the North end and one across the river at Birkenhead. I therefore applied for, and got, a job as a canteen assistant in the North end canteen.

Very soon some of the problems of participative research began to show up. I was easily accepted by the other canteen ladies who were a pleasant and hard working bunch, but it was quite out of character for a canteen girl to want to talk to the dockers about their work and their attitudes to it. This meant I obtained a great deal of information about the problems of running a dock canteen, but little about the problems of being a docker. When I moved to the next canteen at the South end of the dock estate, I decided I must choose a different role. I told the new canteen staff that I was writing a book about the docks and the dockers and wanted to experience the dock situation first hand. This worked splendidly with the dockers who were pleased and flattered to talk about their work, but was not a success with the canteen hands who were suspicious at having a stranger in their midst. They asked, "Why would anyone want to be an assistant in a dock canteen?" With the third canteen in Birkenhead, I tried yet another role, this one closer to the truth. I said I was a student trying to earn some money and also carrying out a project on life on the docks. This new role presented few problems with either the dockers or the canteen assistants.

These experiences illustrate some of the problems of participant observation, both concealed and open. How can the researcher make herself acceptable to the groups she wishes to study and how can she choose a role that will enable her to collect the data she needs? It also raises the question of whether this approach is even ethical. Is it acceptable to collect research data surreptitiously or should the researcher always come clean and tell all the groups she works with what she is doing and why she would like their permission and help in collecting the data? However, the participant observation approach proved excellent in collecting rich data which might otherwise have been inaccessible. At the same time other members of the research team were collecting data using face-to-face interviewing. The two methods enabled the collection of both data that could be treated statistically and observational data, much of which provided an explanation for the replies to the questionnaire.

In this early research insights into stevedoring problems and relationships were discussed with Dock Labour Board Management, with the trade union and with individual dockers. The dockers, as a group, were not involved in solving the problems.

Working Underground

My next major project was also in a "tough, masculine" industry and again the subject was industrial relations, this time in coal mining. The question was: "Why were some industrial relations between workers and management so bad in this industry and how could these be improved?" The research team decided to use questionnaire data to answer these questions but once again it turned into a project also involving participant observation. The British National Coal Board, who sponsored the research, asked the Liverpool research team to study two coal mines, one of which had frequent strikes while the other was strike free. The strike prone mine was to be my responsibility.

It immediately became clear to me that it was pointless interviewing miners on the surface. There would be no way of getting an understanding of the nature of their lives underground. I therefore asked to see each miner in the sample at their underground place of work. This immediately created some interesting challenges for the researcher. My pit, called Maypole, was old and it was necessary to walk for about a mile before the working coal faces were reached. This was not easy, as a low roof meant that one walked bent double on the roadways that took the coal from the coal faces to the pit bottom. Many of the faces were so low that it was necessary to crawl along them and many miners had to be interviewed with the interviewer and the interviewee lying on their backs, side by side and in the dark, except for the lamp we each had in our helmets. The pit was also a dangerous one. There were many accidents, and 50 years earlier there had been a bad explosion, due to methane gas becoming ignited, in which almost all the workers had been killed. So Maypole had a bad reputation and many miners tried to avoid working there (Scott, Mumford, McGivering and Kirkby, 1963)

The miners were at first startled and then intrigued to have a woman underground but all were very pleasant and friendly. Once again, however, some of the problems of being an action researcher began to surface, especially those of easy acceptance into a new group. Interestingly, the first was language, both understanding this and its nature. It took me about three months to became familiar with the local dialect and with some of the local words and phrases. The miners constantly referred to "playing" and it was some time before I realised that this meant being absent from work. They also all used very colourful, and unprintable, words when dealing with awkward machinery. However, some years ago, if not today, this was masculine language and not to be used when women were around. This presented them with a problem. What were they to do if a woman were underground? For a while all bad language stopped, a situation that was unsustainable. Then the underground telephone system was used as a means of relaying a warning: "Watch out she is coming this way."

At this point I decided I had to do something about the situation myself. The result was an effective socio-technical solution. Miners, to stay alive underground, have to have air and most collieries[1] have complex circulation systems in which air is blown into the pit down one air shaft, travels round the different underground workings and emerges through a second air shaft. Anyone travelling underground moves with the air and so smell is very strong and penetrating. My strategy was to cover myself with Chanel Number Nine perfume before going underground. This meant that the miners were forewarned of my presence through a sharp blast of perfume. I followed behind the perfume.

My next problem was an ethical one and not so easily solved. Miners work on three shifts—day, evening and night—and I had to interview a number of miners on each shift. The night shift was very different from the day. It was manned mainly by relatively young boys on their first underground job. The coal face moved forward as coal was cut, placed first on a moving assembly line then in bogies[2] and taken to the pit bottom where it was transported up to the surface. As this advance took place, the old worked face had to be packed with dirt and the pit props removed so that the roof could fall in. The dirt prevented subsidence that would have damaged buildings on the ground above. Apart from the young lads, there were few men underground and little supervision. This meant that men could, from time to time, have a short sleep unnoticed before resuming their work. One night when I was walking along the exit roadways with one of the miners, I asked why there were so few men around. His startling and frightening reply was that some would be asleep, others would be having a smoke. This was in a pit that was full of gas and had blown up in the past killing everyone underground. I wondered, first, if he was joking and next what I should do about this information.

Given the bad relations between men and management, passing the information on to supervision would be seen as a betrayal of confidence and could lead to the termination of the research. I thought long and hard and decided that I must tell management, there was too great a risk of a very bad accident. And so the next morning I went to see the pit underground manager and told him that I believed some of the men were smoking on the night shift. He did not treat the information as a joke but said he had had suspicions about this for some time. I explained that my telling him of the problem would jeopardise my personal relations with the men and asked if he could keep me out of the situation. He responded, very cleverly, by sending himself an anonymous note saying 'men are smoking at the coal face, you must do something'. After this episode very careful searches were carried out on all the men before they went underground.

In this project, as in the docks, research data were fed back to, and discussed with, management and the trade unions, but the miners as a group did not contribute to the solution of underground production and industrial relations problems.

These early examples of participant observation research were very valuable learning experiences for me that carried over to my subsequent action research on information systems. They taught me that in depth research is unlikely to be free from ethical, political and practical problems, all of which must be handled with skill if the researcher wants to remain in the research situation. They also taught me that this kind of research is extremely tiring. The researcher not only has to collect the research data for later analysis, she has to write it up in the evening after work so that she maintains an accurate record of events. But I also learned what a valuable experience it was in providing the researcher with a comprehensive and detailed understanding of the activities, attitudes, and emotions of the group that is being studied. The researcher learns what is important and meaningful to the group she is studying.

EXAMPLES OF ACTION RESEARCH WITH COMPUTER-BASED SYSTEMS

In the 1960s some mysterious machines called computers began to be used by industry and commerce, and my research group at Liverpool University was asked by an international group called the European Productivity Association if they would participate in a joint project with a number of European countries. These included France, Germany, Italy, Norway, Denmark and Sweden. The aim of the research was to establish how computers would affect industry. For example, on the positive side, would they make industry more productive and efficient? And on the negative side, would they displace large numbers of staff? This led to my first computer projects in a bank and in a manufacturer of cattle foods (Mumford and Banks, 1967).

I had now become familiar with the work of the Tavistock Institute of Human Relations and greatly admired its philosophy. This was to improve the quality of working life of people on the shop floor and to persuade employers to take seriously the need to help employees to acquire more skills and take more responsibility. In the late 1970s, I was invited to join the International Quality of Working Life Committee, a group dedicated to spreading the QWL message around the globe. The dominant part of my role now changed from that of participant observer to the more action-oriented one of group facilitator. It was my job to help technical and user groups introducing new computer technology to design systems that brought technical, organizational and social improvements to the work situation. My academic role was to use the knowledge derived from this to test out old theory and develop new.

Groups Can Solve Their Own Problems

A learning experience that confirmed the value of action research came from an unexpected user initiative in a project with a British company, Turner's Asbestos Cement, which made products for the construction industry. The TAC systems analysts were anxious to change the firm's sales office from a batch to a terminal-based system for company accounts. They asked for help saying that they wanted to associate good organizational and job design with the new technical system. I made a survey of job satisfaction in the sales office and discussed the results with all the clerks, bringing them together in small groups. At these meetings a large number of organizational problems emerged and I suggested to the clerks that they should think about how these might be solved. I then forgot about this request and fed back the results of the survey to members of the technical design group. They then designed what they thought was a good socio-technical system, called a meeting of all the clerks, described their proposed system, and sat back waiting for approval. Instead, one of the senior clerks stood up, thanked them for their presentation and said he and his colleagues had also been designing a new work structure for the department, which he would like to present. He then produced an excellent blueprint for a form of work organization that solved most of the office's efficiency and job satisfaction problems.

It was the clerks' solution that was implemented, and I learned my most important lesson about action research and participation. This is never to underestimate a group's abilities. People at any level in a company, if given the opportunity and some help, can successfully play a major role in designing their own work systems. I have used a participative approach ever since (Mumford, 1995).

A Successful Project

Another project that reinforced my belief in user participation in systems design took place soon after the Turners Asbestos experience. Rolls Royce Aerospace made aircraft engines and was a large and flourishing company. It had a purchase invoice department which dealt with the invoices coming in from companies supplying goods and services. This department had an elderly, low morale workforce with little motivation to work efficiently. It was shunned by young people who refused to work in a place they regarded as a graveyard. Rolls-Royce had decided to computerise the clerical processes in this department in an effort to improve efficiency. As I had lectured the Rolls Royce systems group on participative design on a number of occasions, the MIS manager decided to try a participative approach with the new system (Mumford and Henshall, 1979).

A user design group was created with representatives from each section of the department together with the systems analyst responsible for the project. I acted as facilitator to the group, and one of the senior clerks was chosen by the members as their Chairman. At the same time the senior Purchase Invoice Manager, the Head of Management Services, the Personnel Manager and the company's senior Trade Union Official agreed to form a steering committee. The design group began its work by analyzing the Purchase Invoice Department's problems and needs. All clerks in the department were asked to make a written note of their most pressing work problems and to complete a job satisfaction questionnaire. Members of the design group then held small group meetings with their constituents to consider more deeply the reasons for these efficiency and job satisfaction difficulties and to discuss possible solutions. Gradually, the work changes required in the department became clear and were documented as important objectives for the new system. These were discussed with the Steering Committee and approved by them.

The systems analyst and his group accepted the task of creating a technical system that would assist the achievement of these objectives, and the design group turned its attention to identifying three alternative organizational structures that would help to secure the required improvements. Two of these were based on the socio-technical approach of multi-skilled work teams, each responsible for a relatively self-contained aspect of the Department's work.

ETHICS

Stage 1 - Diagnosis of needs

Step 1. Why do we need to change? Discuss existing problems and the improvement offered by new technology.

Step 2. What are the system boundaries? Where do our design responsibilities begin and end?

Step 3. Make a description of the existing work system. It is important to understand how this operates before introducing a replacement or modification

Step 4. Define key objectives. Why does this department or function exist, what should it be doing?

Step 5. Define key tasks and information needs. What are the tasks that must be completed irrespective of how the department is organised or the technology it uses? What information is required to complete these tasks?

Step 6. Measure pre-change job satisfaction. Here a questionnaire is used based on current theories of job satisfaction together with the pattern variables of Talcott Parsons. The design group gives this to everyone likely to be affected by the change.

Step 7. Measure efficiency. A second questionnaire based on the control theories of Stafford Beer is also given to everyone likely to be affected.

Step 8. Assess what is likely to change in the research situation in the future. Design must be for the future as well as the present.

Stage 2 - Setting objectives
Clear efficiency, job satisfaction and future change objectives are now set for the new system.

Stage 3 - Identifying solutions
The design group now identifies and discusses in detail a range of alternative organizational and technical design options. Socio-technical solutions will be included in this, but these are not necessarily the ones that are finally selected by the design group. At this stage the design options are likely to be discussed with the Steering Group.

Stage 4 - Choice and implementation of solution

Stage 5 - Follow-up evaluation

Stage 6 - Reports for the company and academic articles describing the theory and practice of the research.

After discussion with the Steering Committee and a meeting with all the clerks in the department chaired by the senior trade union official, an organizational structure was selected in which teams of clerks would look after all the procedures and personal relations for specific groups of suppliers. Clerks in these teams would aim to become multi-skilled, and a time period of two years would be required to achieve this. A number of clerks in the department saw this new structure as too demanding and asked if they could remain on routine work. A service centre was therefore created to handle routine processes such as dealing with the circulation of mail. It was hoped that this would be a temporary structure, with all clerks eventually becoming multi-skilled.

This new structure transformed the department from a low morale group shunned by young employees, to a motivated and knowledgeable group that became of great interest to those departments in Rolls Royce seeking flexible and knowledgeable staff. Here was yet another example of where a socio-technical approach had led to more freedom in decision making and choice. This, in turn, led to more freedom in work by providing opportunities for responsibility, learning and greater control and autonomy (Mumford, 1996; Mumford and Henshall, 1979). It also led to an understanding that researchers must follow action research projects through to the implementation stage and afterwards. When I returned to check the working of the new system 12 months later I found the departmental manager showing mixed reactions. He was delighted with the operation of the new system but claimed that its success was causing him to lose his best staff. Multi-skilled clerks were new to Rolls Royce and very attractive to other departments. There were attempts to lure these away from Purchase Invoice by offering higher grades and salaries. The manager had to provide compensations which would prevent this happening.

Action Research in the Digital Equipment Corporation

These two examples from the 1970s are representative of many participative projects carried out at the lower levels of organizations which included industrial firms, banks, hospitals, the British Civil Service and the armed forces. In the 1980s and 1990s, what is now called the ETHICS approach began to move up the organizational hierarchy. In the early 1980s a major computer manufacturer, the Digital Equipment Corporation, used participative design to create

XSEL, one of its first expert systems. This was intended to assist sales offices throughout the world with computer configuration. The challenge of the project was the size of the user group and the fact that it was located in many different countries (Mumford and Macdonald, 1989).

When computers are manufactured numerous parts have to be brought together and assembled and, because there are so many, some can be left out, or assembled incorrectly. Because of this a customer who receives a new tailor-made machine may find that it does not work, thereby causing a serious deterioration in the relations between customer and supplier. Digital's attempts to solve this problem had failed and the company believed that an expert system, acting as a electronic aide memoire, could be the answer. In the early 1980s an expert system called XCON was built and installed in Digital's manufacturing plants. This provided a graphic display of how different parts should fit together. It was very successful and was welcomed by the engineers building the computers. Unfortunately, it did not solve the configuring problem completely because this originated in the sales offices. Each salesperson had to detail all the parts in a system ordered by a customer: firstly, to give the customer an accurate estimate of how much the machine and peripherals would cost, and secondly, to send a specification to the manufacturing plant stating exactly what the customer wanted. Few of the sales staff were engineers and the specifications were often inaccurate and caused mistakes in assembly that the manufacturing staff could not identify. These configuring errors caused Digital losses of millions of dollars a year. As a result an expert system for use by sales staff was developed (XSEL). In its mature state XSEL contained 15,000 configuring rules.

XSEL was designed with considerable user participation. The design group contained both technical experts and members of the sales force, and met regularly during the design and implementation stages and for some time after the system was installed.

In addition to the group discussions, two questionnaires were used to assess job satisfaction and efficiency needs, with the analysis of efficiency needs based on Stafford Beer's Viable System Model (Beer, 1989). Design was an iterative process, with the sales force specifying their information needs, the technical members building an embryonic system for them to test, and this process continuing as XSEL grew until the system was regarded as ready to hand over to the sales offices for day-to-day use.

Although only some of the sales people could directly participate in the design process meetings and discussions, Digital kept all the sales offices informed of what was happening through its electronic mail system. Regular reports of progress were sent out and when there were arguments over strategy that could not be easily resolved, the sales offices were consulted. Participation meant that when XSEL was ready for implementation, the sales offices were enthusiastic and very willing to use it. The system was non-threatening, people would not lose their jobs because of it, and it would prevent salespeople from making embarrassing and costly mistakes.

The Digital project was, however, yet another example of the need for action researchers to stay with projects after implementation and to study consequences. Despite the initial enthusiasm the system gradually ceased to be used and the configuration errors increased in number again. The problem was both motivational and technical. Over time the sales force felt that there were few benefits in using XSEL. It added an extra administrative task to their workload when they wanted to focus on "selling." It was also slow. A salesperson could do an imprecise configuration in his or her head faster than XSEL could work through its 15,000 rules.

The Digital project is an example of a socio-technical approach directed at developing new software, not restructuring a department. It did change the salespeople's individual work responsibilities but not in a manner they regarded as improvement. The participative aspect of socio-technical design was successful but the final product that emerged was not. Nevertheless, the experience of using a socio-technical approach convinced Digital that this was the way to proceed in the future. They produced a set of guidelines for managing change based on socio-technical design principles that were used to manage subsequent projects in other areas (Mumford and MacDonald, 1989).

Participative Design for
Management Information Systems

Projects in the 1990s were almost all concerned with using a socio-technical approach to assist managers to select and shape management information systems to meet their particular needs. Firms participating in these exercises included KLM, DutchTelecom, a Dutch company making power tools and a British hospital. The participating groups were now usually senior staff such as managers although the group in the British hospital consisted of nurses. Managers were just as keen to participate in designing the systems they would eventually use as lower level staff but the design process now had to be sped up to meet the limited time they had to devote to the exercise. A shortened version of the ETHICS method was developed called QUICKethics which could be carried out in days rather than weeks.

The focus of Quickethics was to ensure that managers obtained what they considered necessary and relevant information from the new system. They were less interested in reorganizing their jobs and work situations. An important action research lesson here was that the information identification exercise needed to be carried out at intervals after implementation. The manager's information needs would change over time and the technical system would rapidly become obsolete (Mumford, 1997).

In the 1990s the interest in socio-technical design began to wane. Industry moved into harder times, there was plenty of labour available and the objective of most major organizations was to cut costs. New techniques arrived such as Business Process Reengineering which offered to reduce the labour force and improve the share price. These, for a while, were implemented with enthusiasm. Consideration for the quality of work life and personal development of employees faded away to be replaced by downsizing and outsourcing. It is only just starting to return.

THE ETHICS APPROACH TO ACTION RESEARCH

The case studies demonstrate the application of the ETHICS approach to action research. The method developed slowly over time as the author worked with different groups and began to understand how they could be helped to diagnose their needs and develop solutions. ETHICS is shown in its totality on page 58, but its use varied according to the demands and needs of particular situations. Most groups did, however, work through the different stages, usually requiring a time period of several months to complete these. Design group meetings were normally held at two weekly intervals to give members of the group time for considered thought and to ensure their normal work commitments were not disturbed.

A contracted version of the method called QUICKethics was used when managers were defining their information needs. A great deal of personal information was collected through face-to-face interviews and discussion with individual managers, followed by two one-day meetings at which the group of managers identified and agreed upon their priority information needs. The reorganization of work was not part of the agenda (Mumford, 1995).

WHAT DOES ACTION RESEARCH INVOLVE?

Action research is a complex process that requires academic and social skills. Here are some of the factors a researcher needs to be aware of. Lessons that came out of the stevedoring and coal mining research were principally concerned with the importance of establishing and maintaining good personal relationships with the project groups. At the beginning of both research projects, I had to establish my credibility and acceptance with all levels of management, the trade unions and the dockers and coal miners. A good strategy with the latter two groups was to establish who the informal leaders were and approach them first. If I was acceptable to these individuals, the other dockers and coal face workers would also accept me. As the research progressed I had the even more difficult problem of maintaining good relationships with groups who were effectively at war with each other. This required being very careful to maintain confidentiality by not discussing each group with the others until the end of the project, and also being very careful to ensure that confidential information was never allowed to leak from one group to another through any indiscretion of mine. In both situations being a woman was a great advantage, as a man would have been seen as a possible management spy and not allowed to carry out the research.

The steps set out below represent the ideal situation. In practice, however, the realities of real-world situations may mean that some are easier to accomplish than others.

Planning the Research

The first, essential, part of any research project is deciding on, and clarifying, the subject for study. Next comes identifying appropriate theory that can provide an intellectual basis for the proposed research area and choosing research methods that will enable this theory to be tested. Also, and most important, is finding an organization that will welcome the research and obtaining the funding required for the research to be carried out.

None of these activities is easy. Ideally, the researcher has a considerable interest in the subject chosen for study, believes it to be important, and hopes that it is an area where it will be possible to make a contribution to knowledge. Unfortunately, compromises may be necessary. Her choice of subject may be influenced by what the organization that has agreed to the research may consider important or, if she is working for a higher degree, by the interests of her university or supervisor.

Her personal values may now influence how she proceeds. She may not wish to undertake research which will help one group in the research situation but damage the interests of others. She may also reject it if it requires undercover research methods such as concealed participant observation where, like the spy, her interests and motives are not made public . The action researcher will also want to protect her academic neutrality. She is, ideally, dedicated to the pursuit of knowledge in an ethical manner. This means that she is there to assist, not damage, the interests of the group or groups that she is studying.

Using Theory

Good research, especially if it is for a higher degree, requires good theory. In this way it can make an important contribution to knowledge. If the subject area has already been studied by others, then some of this theory will exist and can usefully be retested in the project. Hopefully the researcher, either at the start of the project or as it proceeds, will be able to develop other theories and hypotheses which will throw new light on the subject that is being investigated. If the research area is new and unresearched, then little theory may exist. Some researchers may now wish to use a "grounded" theory approach in which theory emerges as the researcher gains greater knowledge and understanding. This approach can be valuable as it generates and formalises new knowledge.

Although the principal objective of having a theoretical basis to action research is to create new knowledge, researchers following a socio-technical approach will want to test out the validity of the new theory through applying it in practice. Or they may use a reverse approach in which practice is studied to identify the theory behind it. An important research principle for members of the Tavistock group was that practice should lead to better theory and theory should improve practice (Trist, 1993).

While the creation of new knowledge is a principle objective of theory, the action researcher will want to use theory in a number of different ways. Certain theories will guide her behaviour in her relationships with the project group. I have always found theories derived from the notions of homeostasis, complexity, and entropy to be of value. Homeostasis, the tendency of living things to maintain a state of equilibrium in the face of changing conditions, can be a personal goal for a facilitator (Beer, 1972). It is an important part of her role to ensure that a project group is able and motivated to complete its design task without anxiety and trauma. One way of doing this is to act as an effective communicator, keeping external groups informed of what the group is doing, and how it is progressing so that misunderstandings can not occur. At the same time she has to keep the project group informed of how the outside world is reacting to its activities and protect it from misunderstandings. Behavioural science theory relating to group processes is always of value as are the cybernetic notions of requisite variety and positive and negative feedback.

My approach to participative design is to use a democratically selected and representative design group as the principal information collectors and decision takers, but at two points in the research to ask them to carry out a survey of all their constituents. The first of these is a questionnaire on job satisfaction for which Talcott Parson's pattern variables are the theoretical base (Parsons, 1951). The second is an analysis of individual work problems and a simplified version of Stafford Beer's viable system model is now used. (Beer, 1989) Much of the theory that emerged from the research described in this chapter was concerned with the dilemmas, problems and advantages of using participation as a design tool for managing change in complex organizational settings.

The theory that is selected for the project will, of course, affect the methods that are used to test it. Researchers who have the necessary staffing resources and do not have tight deadlines should be encouraged to try using a number of different methods simultaneously. Participant observation can be accompanied by in-depth interviews, the collection of statistics and the analysis of relevant documents. All of these will provide valuable comparative data that can help confirm or challenge the selected hypotheses (Mumford, 1999).

Carrying Out the Research

Action research usually starts either by the researcher asking a company, or other relevant organization, if they will allow her to do the research. Alternatively, if the researcher, or her group, is an established one the company may approach her and ask for assistance. Either of these approaches requires careful thought and preparation. The researcher first needs to talk to key people in the research situation. These will usually be senior management, user department managers, and if the project relates to new technology, a sample of future users. If the users do not want to play an active role, the project is best forgotten. In the 1970s and 1980s many British and European companies had powerful trade unions and their agreement to a project was as important as that of management. If the unions did not want the project it could not take place.

Negotiating Entry and Initial Contacts

A very important aspect of placing the research on sound footing is to ensure that both the researcher and all contacts in the company have a clear, specific and agreed knowledge of what is to take place. There should be no ambiguity or uncertainty. This means that a formal "action" document should be created with a precise specification of processes, objectives and outputs. This should be signed by both management and the researcher, and given to all interested parties. The reality of most research is that what actually happens in the project may deviate from what is written in the document and final objectives may differ from initial ones. But the existence of a starting document will make it easier to discuss why changes have occurred and the reasons for these (Neumann, 1997).

An important early decision for the researcher is to decide on her role in the research situation. Is she going to be an active or a passive participant? This decision will depend on both the nature of the project and her personal skills and role preferences. A passive role will mean that although she will be a keen observer and recorder of the changes that are taking place, she will not intervene in, or comment on, the events she is observing. An active role means that she does intervene in the change situation, but in a positive and helpful manner. One way of doing this is to act as a change facilitator, helping the user group to identify problems, think out strategies, and decide on the nature of future change. Being a facilitator means helping the group to make decisions but doing this in a way that does not influence the members thinking or discussions.

Creating the Design Group

As action research requires group involvement careful thought must be given to the nature of the problem-solving approach that is to be used. This requires that some thought is given to the structure of participation, the content of participation and the process of participation. The structure of participation relates to who is in the problem-solving group and how its work is organised. Some project groups will be chosen by management, others will emerge through volunteers. In the ideal situation they will be elected by the members of the department where the project is taking place and will contain representatives from both sexes, different age groups and different jobs. I have always used this approach when helping users to design new computer systems. At different points in the design process, the entire department should be invited to express an opinion on how the project should proceed and on the required end result. This can be done through meetings and questionnaires.

The content of participation is the nature of the issues about which the group can take decisions. It is useful to have some clear boundaries here so that the group knows which decisions come within its brief and which are excluded. The process of participation involves the acquisition of knowledge so that decisions are taken from an increasingly informed position. It involves learning, the development of effective working relationships, the setting and achieving of goals, and the implementation of solutions.

The role of the facilitator is to help the user group to manage the project. This requires familiarising them with the method or methods that are to be used and helping them to mould these to their own requirements. She must also assist them in resolving conflicts of interest or approach and make sure that important aspects of the design programme are not forgotten. In no circumstances must she make decisions for the group or persuade them that certain things should or should not be done. They are the decision makers. The facilitator's role is to encourage the design group members to systematically analyze their own problems and needs, and to arrive at a solution that addresses the problems and meets the needs. In the author's experience the facilitator also needs to keep the group moving towards the completion of the task. She must ensure that the members feel they are making progress and not slowing down or coming to a halt because of disagreements or difficult problems. And she will also make sure that they always keep the project steering group informed of how they are progressing.

She will also have an important communication role. A factor leading to successful action research is ensuring that all interested parties in a change are always kept informed of what is taking place. This can be done through individual and group meetings, through electronic mail and information-providing documents such as regular newsletters. No one in the situation should be able to say: "I don't know what is going on." Communication is often most successful when it emanates from a responsible central source and is then distributed around the organization. The facilitator can act as this effective central source.

Maintaining Relationships

Staying in requires developing and maintaining over time the good relations that have been created at the 'getting in' stage. The researcher now has the difficult problem of needing to be accepted by a number of different groups, some of whom may not be on good terms with each other. These groups will include middle management, any specialist groups that are involved such as systems designers, the participants in the project, and if the firm is unionised, the local trade union officials. It is at this point that the action researcher has to make a very important decision. Who in this complex situation is her principal client, the group whose interests she will place first?

In those projects that involved lower level clerical staff, it was these who were the principal focus of the author's attention and whose interests and well being were her first priority. She assumed that in helping these groups to secure improved job satisfaction and improved quality of working life, there would also be important spin offs for the other groups in the situation, but her principal focus was always the clerical group.

Action researchers must recognize that they are operating in volatile political situations where there may be different, even hidden, agendas. It is important to be aware of the internal politics, but at the same time to keep detached from them. It is not unusual for some individuals and groups, knowing that the researcher has access to senior management in the company, to try and make use of her for their own ends. She must be aware of this, retain her neutrality and avoid being manipulated.

One excellent way of avoiding problems is to ensure that all lines of communication to different groups are kept open and work effectively. She should ensure that everyone in the research situation, even if only indirectly involved in the research, continues to understand what she and the project group are doing and why. This requires the setting up of regular meetings throughout the project at which progress and new developments can be discussed. It is particularly important to keep senior management informed so that their support of the project does not slacken. Action research is a dynamic process and, although methods and objectives are carefully explained at the start of the project, unexpected events together with an increased knowledge of the problem situation may cause these to be revised.

The action research route is not an easy one and can be very difficult indeed. The researcher will always place the interests of her project group first but, merely by involving them, she may cause unanticipated problems. For example, in one project in Rolls Royce Aeroengines the democratically elected design group was not very expert in communicating with its constituents, the other clerks in the department. One member of the design group had always received a cake from her colleagues on her birthday. The year of the project her colleagues became unsure and suspicious of what she was doing and she did not receive the cake. She was very upset.   The researcher learned from this that although most groups can become good designers, it is much harder for them to become good communicators. It may be necessary for the researcher to take on the communication role.

What are the problems that researchers can experience once the research is underway? The very worst, which fortunately I have never experienced, is being asked to leave the research situation. This is most likely to happen through deterioration in relationships with an important group. Top management may feel that she is wasting people's time without achieving good results. Computer specialists may think her presence is slowing the project down. Her project group may become distrustful of her, believing that she is a secret agent for management. However, sometimes being encouraged to leave is a sign of success. This happens when the project group comes to believe it has the knowledge and experience to carry on on its own. It needs to show that it can do this without assistance. Now is the time for the researcher to retire with a good grace and the satisfaction of knowing the project has been a success.

Another difficulty I have experienced is a project group, often clerical, becoming suspicious of management's motives in permitting the project to take place. The members believe that there may be a hidden agenda. In the Rolls Royce Aeroengine project this did happen and the only solution was to ask a senior manager to come and talk to the clerks and assure them that he and his colleagues were enthusiastic about the project and were not going to interfere in the processes or outcomes.

The fact that an action researcher spends a considerable period of time with a firm means that she gains a reasonably good idea of what is happening in the company, but she may not know everything. A project in the Plastics Division of ICI provides a good example. Here many weeks were spent with a group of sales clerks helping them to develop a new form of work organization prior to the arrival of a new computer system. Just as the project was finished to general approval ICI announced it was closing the division down. There would be no new system and no employing firm. This was a disaster.

A decision that needs to be made toward the end of most projects is who is going to take responsibility for implementing the new system of work. Is this going to be the departmental manager, a newly formed implementation group or will the existing project group change its role from design to implementation? Also will the new system be phased in slowly or brought in overnight with a big bang? Training may be a factor here. If the new work roles require time for this, then the duration of the changeover period will be affected. The move from routine to multi-skilled roles in Rolls Royce required a period of two years before all clerks who wished to become multi-skilled were able to do so. This meant the new system had to be implemented slowly.

Action research always requires some careful assessment of results once the project is completed. A variety of methods and respondents can be helpful here. But it is easy for the researcher to be led astray. For example, in one evaluation of a nurse management hospital system, all the technical staff and administrators were convinced of the system's success. It was only when the nurses were interviewed that the very large flaws it contained emerged.

Leaving the Research Situation

Leaving the research situation has its own particular problems. One of the most important goals of the action researcher should be to hand knowledge over to the group she has been working with so that, when she leaves, the group can continue solving its problems. They now have the knowledge, experience and motivation to do this. It is a good sign when a project group starts suggesting to the researcher that it is time for her to depart because they can now handle change themselves. It is not always easy to achieve this. The opposite can occur and the group become over dependent. This happens if the researcher has played too interventionist a role, proffering too much advice and steering the group in a particular direction. I had an important learning experience in this respect when collaborating with other socio-technical practitioners. It seemed that their project groups always ended up with conventional socio-technical solutions including multi-skilling and team decision making. Although, theoretically, this was a very desirable result I came to believe that the way it was achieved had disadvantages. It seemed that these socio-technical experts were having too great an influence on the thinking of the project group. It was their solutions, not the groups' that were being implemented.

Ideally, the project group should be able to continue with its critical thinking once the researcher has gone. This happened in the Digital Equipment Corporation project where the group continued meeting for several years. In contrast, in one of the ICI projects, a group of secretaries became very frustrated once the project ended. Their managers did not realise that they had now become expert at identifying problems and planning for change. The old hierarchical manager-secretary relationship was resumed with great damage to secretarial morale.

The conclusion of an action research project can present the researcher with ethical problems. The members of the project group will have become her friends and told her many things in confidence. Has she any right to publicise these confidences in the learned articles which she subsequently writes, even though they may throw an important light on the culture of the group? One way around this is for the researcher to show the group what she has written and get their approval. Another more revolutionary approach is to get the members of the group to write part or all of the article themselves. The author has tried this occasionally but found that academic journals did not approve of this democratic approach. They were unwilling to publish. The researcher can also experience management censorship. This happened to me when carrying out research for a British government department handling employment issues. The initiators of the research wanted only results that showed them in a good light to be published, anything critical must be deleted. I decided not to undertake any further research for this group. Other research was for the Ministry of Defence and this involved signing the Official Secrets Act. Happily this did not act as a publication restriction as the Ministry was interested only in banning information that could have been of use to an enemy.

Usually at this stage the researcher has the labour of writing practical reports of interest to the management of the participating organizations and also articles for learned journals that will further her personal career. It is not easy to combine both objectives in the same publication but occasionally it can be done. Some firms are interested in theoretical conclusions while increasingly journals want to test the validity of theory by assessing the practice that led to its development.

Follow-Up Studies

These are essential in an action research project. It is important both to review the project as a learning experience and to check how well original objectives have been attained. Most projects experience slippage—the post-change situation alters as time passes, new staff arrive and circumstances change. In Rolls Royce Aeroengines an unexpected development was the attraction of the now multi-skilled staff for other departments in the company. Attempts were made to lure them away much to the annoyance of the departmental manager.

A useful post-change exercise carried out by the author is to once again ask all staff in the department to complete the job satisfaction and problem identification questionnaires and discuss the results in small groups. This can provide evidence of whether the project objectives, which will have included more job satisfaction and problem elimination or reduction, have been completely or only partially achieved.

SOCIAL SKILLS—REQUIRED AND ACHIEVED

Action research both requires social skills and provides an opportunity for acquiring these. The research methods used will influence which of these needs to be given priority. Participant observation which I used extensively in my early projects is always worth considering as a means for obtaining an in-depth knowledge of the pressures and problems a particular group is experiencing. The researcher, if only for a short time, can join in the group's work activities, participate in its social and work relationships, and experience its stresses and strains. However it is not always possible to do this. In many situations I found that I did not have the skills and knowledge necessary to join in the group's work activities. My active presence would just have been a nuisance. The next most useful alternative to participant observation is non-participant observation. Gaining an understanding of what the group has to do by sitting with them or near them as an observer and, when convenient, talking to them about the tasks they are required to do and the problems they experience in doing these. Face-to-face interviewing also requires considerable social skill, especially if the interviewees are senior managers. They need to be put at their ease and made to feel that talking to the researcher will not be a waste of their time. However most people like talking about their work and the researcher may be the first person who has ever asked them to describe in detail their roles and responsibilities. We have already discussed the role of the facilitator. In my experience the researcher learns how to be a facilitator by being one. Books can give good advice but real knowledge of how to help a group to solve problems and, at the same time, keep up their momentum and their morale only comes from experience.

Researchers need to acquire 'competence' in the research role. Competence has been defined by the philosopher Gilbert Ryle as 'knowing how' (Ryle, 1949). He argues that in every day life we are much more concerned with people's competencies than with their intellectual brilliance or with their beliefs. Competence implies that a person presented with a problem can think things through logically and get results. Socio-technical competence requires a knowledge of how to achieve social and organizational as well as technical goals. Set out below are some of the competencies that an action researcher is likely to find useful. She will also find that action research is a means for developing these (Mumford, 1999).

Knowledge competence:

The ability of an individual to learn from experience and, as a result, to continually add to personal knowledge. The researcher may not have a great deal of knowledge about the research situation when the research begins; her knowledge will increase as it progresses.

Resource competence:

An understanding of the kinds of personal skills that are required. These will include management, technical, organizational and social skills. She needs to be able to plan her work, understand the technology that is involved and its organizational implications and have the social skills to relate successfully to staff at all levels.

Psychological competence:

An ability to work with, motivate and encourage both internal and external groups, to maintain personal morale in demanding and stressful situations and to persevere with difficult problems.

Organizational competence:

To plan and think strategically about her personal contribution and how this can fit with the needs of the total systems design situation. What can she contribute to helping users and management improve the work environment and what can she contribute to the academic community?

Innovative competence:

An ability to think creatively, to approach new and unexpected problems without prejudice and from different angles.

Ethical competence:

An understanding of and willingness to communicate her personal ethical values where these are relevant to the needs of the project and those concerned with it.

Successful systems design requires a great deal of coordination with different groups working effectively together. The action researcher must understand these needs and, whenever possible, assist groups to work together more effectively. Socio-technical design requires decisions to be taken, or influenced by, the groups most likely to be affected by them. These, according to how the project is organized, will range from software suppliers, outsourcing contractors and external consultants to internal managers at all levels and to users of the system. Critical groups will be external users, or recipients, of the system, especially if they are customers or members of the public. The researcher needs to be aware of these complex networks of relationships, understand their nature and, whenever possible, provide assistance.

CONCLUSIONS

Action research is satisfying but difficult. It requires time, social skills and the ability to transfer personal knowledge to a group that is often inexperienced in the management of change. Action research also requires a knowledge of history. What was the situation before the project started and what happened after it finished? Projects, although initially very successful, may not endure or diffuse. A new manager may arrive who does not understand or approve of the democratic processes which have taken place. Other departments may be jealous of the attention the user area has received and may not want to follow its example. Nonetheless, with industry rapidly becoming both more complex and more international change will be on every manager's agenda. Action research and socio-technical design and are well worthy of consideration as approaches and tools for the future.

REFERENCES

Beck, U. (1992). *Risk Society*, London: Sage.

Beer, S. (1972). *Brain of the Firm.* Wiley: London.

Beer, S. (1989). 'The Viable Systems Model: Its Provenance, Development, Methodology and Pathology in R. Espejo and R.Hardin (eds.) *The Viable System Model* London: Wiley.

Castells, M.(1996). *The Rise of the Network Society.* London: Blackwell.

Department of Trade and Industry (1999). *Work in the Knowledge Driven Economy,* London.

Emery, F. (1993). The Historical Validity of the Norwegian Industrial

Democracy Project in E.Trist and H. Murray (eds.) *The Social Engagement of Social Science. Vol. 11.* Philadelphia: University of Pennsylvania Press.

Giddens (1991). *Modernity and Self-Identity,* London: Polity Press.

Handy, C. (1994). *The Empty Raincoat.* London:Hutchinson.

Leadbetter, C.(2000). Comment, *New Statesman,* 6th March, 26-28.

Mumford, E.(1995). *Effective Systems Design and Requirements Analysis.* London: Macmillan.

Mumford, E.(1996). *System Design: Ethical Tools for Ethical Change.* London: Macmillan.

Mumford,E.(1999). *Dangerous Decisions: ProblemSolving in Tomorrow's World.* New York: KluwerAcademic/Plenum.

Mumford, E.(1997). ETHICS: User led Requirements Anlysis and Business Process Improvement in K.Vargese and S Pileger (eds.) *Human Comfort and Security of Information Systems.* Berlin: Springer.

Mumford, E and Banks,O.(1967). *The Computer and the Clerk.* London: Routledge and Kegan Paul.

Mumford, E. and Henshall, D.(1979). *A Participative Approach to Computer Systems Design.* London: Associated Business Press.

Mumford, E. and Macdonald B. (1989). *XSEL's Progress.* London: Wiley.

Mumford, E. and Pettigrew, A.(1975). *Implementing Strtaegic Decisions.* London. Longman.

Neumann, J. (1997). Negotiating Entry and Contracting in J.E.Neumann, K. Kellner and A. Dawson-Shepherd, *Developing Organizational Consultancy.* London: Routledge.

Parsons,T.(1951). *The Social System.,* London: Routledge and Kegan Paul.

Pettigrew, A.(1973). *The Politics of Organizational Decision Making.* London: Tavistock.

Ryle, G.(1949). *The Concept of Mind,* London: Hutchinson.

Scott,W.(1963). Mumford,E., McGivering, I., Kirkby, J. *Coal and Conflict.* Liverpool: Liverpool University Press.

Simey, T.S. (1954). *The Dock Worker.* Liverpool: Liverpool University Press.

Trist, E. and Murray, H.(1993). *The Social Engagement of Social Science. Vol 11: The Socio-technical Perspective.* Philadelphia: University of Pennsylvania Press.

Whyte. W.F. and Hamilton, E.L.(1965). *Action Research for Management* Illinois: Irwin.

ENDNOTES

1. A colliery is the British word for coal mine. I worked at Maypole Colliery. Most collieries had a number of pits, the name for their individual underground workings.
2. Bogies are tubs or trucks for carrying coal. The coal comes off conveyor belts and goes into tubs. These are then sent up to the surface.
3. ETHICS is an acronym for, Effective Technical and Human Implementation of Computer-based Systems. It is an ethical approach to IT design.

Chapter IV

Reflexive Ethnography in Information Systems Research

Ulrike Schultze
Southern Methodist University, USA

To do fieldwork apparently requires some of the instincts of an exile, for the fieldworker typically arrives at the place of study without much of an introduction and knowing few people, if any. Fieldworkers, it seems, learn to move among strangers while holding themselves in readiness for episodes of embarrassment, affection, misfortune, partial or vague revelation, deceit, confusion, isolation, warmth, adventure, fear, concealment, pleasure, surprise, insult and always possible deportation. Accident and happenstance shapes fieldworkers' studies as much as planning and foresight; numbing routine as much as live theater; impulse as much as rational choice; mistaken judgments as much as accurate ones. This may not be the way fieldwork is reported, but it is the way it is done (Van Maanen, 1988, p. 2).

INTRODUCTION

While I was endeavoring to write up my first piece of ethnographic research, which also happened to be my dissertation, I found the above quote reassuring and inspiring. It was reassuring because it aptly described my experience in the field and therefore gave me the sense that it had been "normal." It was inspiring in that it prompted me to explore ways of reporting my research such that its textual reproduction reflected more accurately how the research had been done. Other texts written by anthropologists on the work of writing ethnography (e.g., Clifford and Marcus, 1986; Kleinman and Copp, 1993; Marcus and Fischer, 1986; Wolf, 1992), as well as research adopting a confessional genre (Van Maanen, 1988) or a vulnerable style of writing (Behar, 1996) led me to experiment with a reflexive mode of representation. Given the centrality of writing in ethnographic research, it quickly became apparent to me that the style of writing has implications not only for what can be said, but also for how the data can be analyzed. Thus, true to Marcus and Fisher's (1986) assertion, my choice of representational genre led me to significant insights about the parallels between my practices of informing as a researcher and the informing practices of the knowledge workers I had studied (Schultze, 2000).

In this book chapter I would like to present reflexive ethnography as a particular type of ethnography and explore the factors that I found to be key in completing it. Myers (1999), citing Sanday (1979) introduced three different types of ethnography to the IS community. He described the *holistic* school of thought, which insists on the ethnographer going native and thereby identifying and empathizing with the social group being studied. The *thick description* approach to ethnography denies the importance of empathy for and identification with the group being investigated, focusing instead on their symbols, that is, language, images, institutions, and behaviors. *Critical ethnography* assumes that the social order is influenced by hidden agendas, power centers and assumptions that repress and hide realities. Such ethnographies seek to uncover what is hidden from plain view.

In contrast to these approaches to ethnography, *reflexive ethnography* is conducted and written up in a way that takes into account the researcher's self in interaction with the object of study (Davies, 1999). It openly acknowledges the role of the ethnographer throughout the fieldwork—from choosing the research topic and field site to writing up the research—and it regards his/her emotional and deeply personal reactions to the field and the method as legitimate data (Kleinman and Copp, 1993). While a confessional style of writing is frequently an indicator of a reflexive ethnography, the two are not synonymous, in that a reflexive ethnography recognizes fieldwork and its subsequent representation as self-presentation and identity construction primarily of the individual researcher (Coffey, 1999; Kondo, 1990), but also of social research and society (Davies, 1999).

In this chapter, I will define ethnography and discuss why it is particularly important in IS research. Then I will discuss the importance of reflexive ethnography specifically and identify five factors that I found to be key in doing a reflexive ethnography: the willingness and ability to become immersed, the availability of a supportive thought community, the tenacity to carry on despite ambiguity, the discipline to write self-reflexive field notes, and the ability to identify ways in which this self-reflexive material can be used to make a substantive contribution to IS research. I conclude this chapter with some thoughts on the benefits of doing reflexive ethnography.

WHAT IS ETHNOGRAPHY?

Ethnography is a research method that relies on first-hand observations made by a researcher immersed over an extended period of time typically in a culture with which he/she is unfamiliar[1] (Agar, 1986; Atkinson and Hammersley, 1994; Hammersley, 1992). The ethnographic method requires the researcher to closely observe, record, and engage in the daily life of people in the field, and then to write about it in descriptive detail (Marcus and Fischer, 1986, p.18). Ethnographers rely on a wide variety of data collection methods ranging from surveys, through tape-recorded interviews, to participant observation and the review of documents. Despite this diversity, personally being in the field and writing fieldnotes that describe personal observations and experiences are the hallmarks of ethnographic research (Amit, 2000; Kleinman and Copp, 1993). While the ethnographer acts as the research instrument, the fieldnotes represent the inscriptions that make the instrument's observations and experiences immutable and mobile[2]. As such, the fieldnotes serve as the basis for data analysis and evidence in the final representation of the research.

The questions with which ethnographers enter the field are generally broad and exploratory (Carspecken, 1996). An example of such questions is 'What is life like for this group of people?' The broad and open-ended nature of ethnographic questions gives 'the material free rein to find its way to central questions and answers' (paraphrased from

Mack, 1971, p. vii). This also implies that the design of an ethnographic study is loose and somewhat messy. However, it leaves room for the emergence of unanticipated findings and impromptu changes to the initial research design based on the ethnographer's insights, questions and ideas generated in the field.

Ethnography is particularly suitable for research where the problem is not clear and complex, and where the phenomenon is embedded in a social system that is poorly understood or even unknown (LeCompte and Schensul, 1999, p. 29). It is conceivable that these conditions exist frequently in the discipline of information systems. The next section therefore looks more closely at why ethnography is important in information systems research.

WHY IS ETHNOGRAPHY IMPORTANT IN INFORMATION SYSTEMS RESEARCH?

The use of ethnographic methods in management and information systems research is increasing. Topics that are being explored include system implementation (Myers, 1994), system design and development (Bucciarelli, 1994; Forsythe, 1993, 1995; Myers and Young, 1997; Suchman, 1993, 1995), organizational change (Barley, 1996; Orlikowski, 1991, 1996; Perlow, 1998, 1999; Prasad, 1993), organizational learning (Orr, 1990, 1996; Pentland, 1995; Spitler and Gallivan, 1999), informing processes (Preston, 1986; Schultze, 2000; Schultze and Boland, 2000, forthcoming) and the automation of work (Wynn, 1991; Zuboff, 1988).

This apparent growth of ethnographic research is laudable given rich insights into the human, social and organizational aspects of information systems that can be gained from ethnographic IS research (Myers, 1999). Such insights might affect the high implementation failure rates in technology projects. For instance, some have argued that a better understanding of work practices is required to improve the success rate of implementation projects (e.g., Button and Harper, 1996; Grudin, 1994) and the organizational transformations anticipated as a result of technological innovation (Suchman, 1995). Work practices represent the "lived work" (Button and Harper, 1996) of people as they interact with others and with material and immaterial things. Since a practice orientation focuses on what people *actually* do, rather than what they *say* they do or what they *ought* to be doing (Pickering, 1992), ethnographic research is particularly well suited to answering the kinds of questions a practice orientation generates (Barley, 1988; Blomberg, Giacomi, Mosher, and Swenton-Wall, 1993; Brown and Duguid, 1991).

The study of work practices is also important in our contemporary climate of ongoing technological advancement, where change is increasingly regarded as the only constant. If change is the order of the day, how can social scientists find enduring relationships and patterns on which to base research findings and recommendations? Work practices, according to Bourdieu's (1992) theory of practice, are mutable but also have enduring characteristics that make them suitable to scientific inquiry. Similarly, in structuration theory there is a notion that actions enable and constrain further action (Giddens, 1987), thus suggesting that practices are somewhat persistent. This persistence makes the study of work practices a worthwhile focus of information systems research and ethnography a particularly useful method, since it is well suited to the inquiry of work practice in their everyday, mundane detail.

Another reason why ethnographic methods are important in IS research is that technology is socially constituted (Orlikowski, 1992; Orlikowski and Robey, 1991). This implies that any inquiry of technology requires a thorough grasp of the structured and structuring influence of social action on the development and use of information technology. This structurational relationship between the social environment and the technology is particularly important in the context of communication technologies, such as e-mail and groupware, where technology, information and social action are inextricably intertwined (Lyytinen and Ngwenyama, 1992).

Having explained why ethnographic methods are important in IS research, it needs to be highlighted that there are usually a whole host of reasons other than the fit between the research question and the methodology that account for a researcher's methodological choice. For instance, there are political or career reasons why researchers pick one method over another[3]. In doing reflexive ethnography, it is necessary to identify the decisions made with respect to topics, research questions, methodology, research design and field site and why these were made, as these impact the researcher and the research (Davies,

1999). I will discuss some of the reasons why I chose the ethnographic method for my dissertation in this next section.

My Reasons for Doing an Ethnographic Dissertation

Throughout my graduate work I found myself fascinated by the social processes through which information is created and the role that information technology plays in this process. I was particularly interested in understanding the difference between hard and soft information[4], and how work practices and technology contribute to the process of hardening and softening. An ethnographic approach seemed the most appropriate for answering questions of this nature, since the definitions of hard and soft information are vague and situated in local contexts. Furthermore, there was little prior research on the "practices of informing" and "knowledge work," the labels that came to capture and define my research focus over time. Other reasons for choosing this mode of research included a desire to learn about American corporations first hand, a supportive dissertation committee and the desire to try this method while I had the luxury of time. I address each of these below.

Being a foreign student in the United States, I wanted to learn first hand what life was like inside an American corporation, especially because I saw myself teaching in an American business school after completing my degree. Even though I had done some interview-based research in a number of American corporations during my Ph.D. course work, I still felt that my understanding of corporate America was somewhat lacking. Doing ethnography, I reasoned, would provide me with valuable insight into American corporate culture.

At the 1990 *IFIP 8.2* Conference, during a panel discussion on relevance versus rigor in IS research (Turner, Bikson, Lyytinen, Mathiassen, and Orlikowski, 1991), Wanda Orlikowski remarked on the institutional barriers in the United States that make interpretive research challenging. She noted that few senior IS faculty have the experience and expertise to supervise qualitative and interpretive dissertations. Having read the transcript of this panel discussion and finding myself in the fortunate position in which I had a dissertation committee that was supportive of qualitative research, as well as a dissertation advisor and a committee member who were very familiar with ethnographic methods, I felt somewhat compelled to take advantage of these apparently rare circumstances.

In the same panel discussion, Wanda Orlikowski remarked that it was very challenging to do qualitative, and especially ethnographic, research as a faculty member, due to the time-consuming nature of data collection (also, Myers, 1999). Since I suspected that I would not get much of a chance to learn and do ethnographic work once I found myself in a tenure track position, I decided to take advantage of the freedom that the Ph.D. program granted me to engage in this time-consuming, high-immersion research.

Thus, in addition to selecting an ethnographic approach because it was the most appropriate for my research question, I was motivated to do ethnography for other reasons. These played an important role in shaping my view of myself as I entered the field and in constructing my identity as an ethnographer once I was in the field. Before describing this fieldwork in more detail, however, I will discuss reflexive ethnography.

WHAT IS REFLEXIVE ETHNOGRAPHY?

"Reflexivity, broadly defined, means a turning back on oneself, a process of self-reference" (Davies, 1999, p. 4). In all science, reflexivity is desirable in order to identify what role the researcher has played in arriving at certain findings or results (Bourdieu and Wacquant, 1992). In ethnographic research, where the ethnographer is the research instrument, a "turning back on oneself" is particularly important. Indeed, the issue of reflexivity lies at the heart of the crisis of representation and legitimation that ethnographic methods have been undergoing in recent years (Denzin, 1997). This crisis represents the lingering tension between the goal ascribed to by most scientists, which is the discovery of objective[5] truths that are independent of the social context and the individual knower, and the subjective, personal and experiential nature of ethnographic data collection, analysis and writing. The authority of the *ethnographer as scientist* depends on his/her ability to convince readers that the phenomena he/she is describing exist outside of the researchers' mind (validity) and that they are observable by others as well (reliability). The *ethnographer as instrument*, in the meantime, is constantly confronted with his/her own subjectivity, including emotional reactions such as fear, joy, likes, dislikes, and breakdowns in understanding and assumptions such as surprise, confusion and consternation.

Personalized accounts of fieldwork – or what van Maanen (1988) labels confessional tales — focus on the ethnographer as instrument. It requires the ethnographer to give a self-revealing and self-reflexive account of the research process. This reflexivity enhances ethnographic authenticity (Golden-Biddle and Locke, 1993) because readers are given insight into who the ethnographer is and what the fieldwork was like. By being able to (almost) step into the ethnographer's shoes and thereby enter the field vicariously, readers can assess the reasonableness and credibility of the ethnographers' interpretation and reconstruction of the field. Behar (1996, p.16) points out that "when readers take the voyage through anthropology's tunnel, it is themselves that they must be able to see in the observer who is serving as their guide".

However, by focusing the ethnography on the research experience, confessional tales tend to deal more with methodological than substantive issues (e.g., Prasad, 1997; Trauth, 1997). As autobiographical details pertaining to the ethnographer form an important part of confessional accounts, this is a genre that blurs the distinction between ethnography and autobiography, as well as between science and fiction (Okely and Callaway, 1992; Reed-Danahay, 1997).

Furthermore, by mixing the public and the private, confessional writing exposes the ethnographer, rendering his/her actions, failings, motivations and assumptions open to public scrutiny and critique. By revealing themselves in this way, ethnographers relinquish authorial power and their "professional armor" (Okely and Callaway, 1992, p.7). They put themselves on a par with their 'subjects' who typically feel exposed and criticized by ethnographic texts (Miles, 1979; Whyte, 1996). Thus, being openly reflexive can undermine the researcher's credibility and authority[6].

However, the criticism that authors writing in a vulnerable style level against themselves also reflects on their community. By juxtaposing their assumptions and practices with those of the foreign culture, ethnographers prompt their readers to do the same, thereby allowing the foreign culture to serve as a mirror in which the reader's own assumptions and practices are reflected. Thus "cultural critique," which implies the use of foreign customs and practices to reflect on the world of the researcher and reader (Marcus and Fischer, 1986), may be achieved. Furthermore, based on Behar's (1996, p. 16) assertion "when you write vulnerably, others respond vulnerably," reflexive texts can be seen as an effective method for achieving the goal of criticality — moving readers to examine their own taken-for-granted assumptions — which Golden-Biddle and Locke (1993) identify as one of the three criteria for a convincing ethnography.

Having outlined what constitutes reflexive ethnography, namely, self-conscious fieldwork and reflexivity in the data analysis and the subsequent representation of the field, I will now outline my dissertation research project, a workplace ethnography. I conducted this study in a very self-reflexive way without being aware of reflexive ethnography as a particular type of ethnographic research.

THE RESEARCH PROJECT: ON THE INFORMING PRACTICES OF KNOWLEDGE WORKERS

The aim of my dissertation research was to explore knowledge work *in situ* in order to develop a practice-based understanding of what creating information entails and what role information technology plays in it. Arguing that knowledge workers looking for a technology to help them improve their information-creating processes represented an optimal field site, I identified a technology to lead me to a situation in which I could study people engaged in creating information and knowledge. The technology I chose as a guide was KnowMor[7], a knowledge management technology that resided on a Lotus Notes platform. My prior experience with research on Lotus Notes influenced this decision. KnowMor was built on the premise that news only constituted information when a group of people agreed that it had relevance to their organization. Its design therefore embodied an informing process of "alert-assess-escalate," which the KnowMor brochures also referred to as the "gatekeeper model." A "gatekeeper" designated a person who either had an interest in a topic area or was deemed an expert in a subject matter. His/her task was to scan continuously the environment so as to be "alerted" to relevant news, to "assess" it, and to "escalate" it, i.e., pass onto others, if it had particular importance to the organization.

US Company, a large Fortune 500 manufacturing firm[8] headquartered in a midwestern state of the United States, was about to embark on a KnowMor implementation project. Their plan was to give the technology to researchers, including scientists and engineers at US Co's R&D facility, as well as marketers, competitive intelligence analysts and senior managers located at the corporate headquarters. For a variety of reasons, including the company's location within driving distance from my home, the early stage of their KnowMor implementation and the diverse composition of the intended user community, I chose US Co as my field site. In 1995, the company recorded sales of $3.6 billion and employed 17,000 people worldwide.

Negotiating access was easier than I had anticipated. Within a week of contacting Jerry Hunt[9], the competitive intelligence analyst who was championing the KnowMor project, I was given permission to conduct my research. I believe that Jerry Hunt's informality and the promise of visibility, legitimacy and prestige that the research activities of a Ph.D. student lent the KnowMor project made gaining access fairly easy.

Employing a technology as a guide to a venue in which to study work practices related to information can be effective in that the introduction of a new information system is typically associated with a certain amount of disruption and change. During times of change, people are forced to engage in more sensemaking of their past, their present, and their future than they do otherwise. Situations of organizational and technical change thus offer themselves as windows of opportunity for studying social phenomena (Zuboff, 1988).

However, relying on a technology as a guide can also be risky due to the high implementation failure rates common among technology projects. In the case of my research, KnowMor was not implemented successfully in US Co. Following the technology was nevertheless effective, albeit in ways that surprised me. Whereas I counted on KnowMor to lead me to study the informing practices of the scientists, engineers and managers who were its intended users in US Company, the knowledge workers it led me to instead were the people promoting and implementing KnowMor. These were three system administrators, three librarians and four competitive intelligence analysts. Only the competitive intelligence analysts were among the intended users of KnowMor.

These three groups were well suited to studying the practices of informing because the production and reproduction of information was central to their responsibilities. The competitive intelligence analysts represented the archetypal gatekeepers as they scanned the business environment, assessed the credibility of sources and the significance of events, and alerted decision makers in the business units to important developments both inside and outside the organization. The librarians' mission was to respond to their customers' information requests. They translated their customers' questions into on-line search commands and retrieved the information their customers needed. The system administrators were responsible for maintaining a stable Lotus Notes environment. This required them to not only configure the hardware and software, but also generate information primarily in the form of documentation that would assure the systems' smooth operations on an ongoing basis.

I was in the field over an eight-month period, from October 1995 to mid-May 1996. Similar to the research method Preston (1986) employed in his research on the use of information technology in informing behavior, I was in US Company Monday through Thursday. My stay amounted to a total of 111 days. Throughout my fieldwork, a typical week started early on a Monday morning and ended late on a Thursday evening. I drove either to HQ-City (a two-hour car ride) where US Company's head office was located, or to its research and development facility at R&D site (a three-hour car ride). During the week I stayed in a motel if I was in HQ-City or with Norma, the leader of the US Co library, if I was at R&D site.

I experienced fieldwork as constant writing (also, Clifford, 1986). While I was physically in a place as an actor on the scene, I was simultaneously suspended above the setting, observing myself and others from a distance and putting into words what was going on. On converting these words into written text, I again found myself in multiple places, i.e., physically in front of my computer with my handwritten notes, and mentally replaying the scene. This heightened sense of self-awareness and reflexivity was captured in my fieldnotes. During the write-up phase of my research, I felt compelled to include myself in my representation of the field. This led me to incorporate my fieldnotes into my manuscripts, and to adopt a confessional genre of writing for some of these manuscripts.

Thus, unintentionally, I had begun to rely on a reflexive approach to ethnography. In reflecting on my experience, I identified several factors that were key in completing a reflexive ethnography. The next section discusses these.

WHAT DOES IT TAKE TO DO A REFLEXIVE ETHNOGRAPHY?

LeCompte and Schensul (1999, p. 170) summarize the characteristics and skills required by ethnographic fieldworkers as follows:

- adventurous, resourceful, enthusiastic, self-motivated, trustworthy, risk-taking, curious, and sociable;
- able to be observant, be communicative, think conceptually, be cognizant of cultural issues and behavior, be reportorial, remember well, separate strict observation from personal bias or opinion, and work well with a team.

This is a useful checklist for anyone considering ethnographic fieldwork in that it highlights the social, conceptual and risky nature of such an endeavor. In reflecting on my own experience in the field, however, the five factors that stand out are (1) the willingness and ability to be immersed, (2) the availability of a supportive thought community outside the field, (3) the tenacity to carry on despite ambiguity, (4) the discipline to write self-reflexive fieldnotes, and (5) the ability to find a way of integrating the self-reflexive material with the "actual" ethnographic material to make a substantive contribution to IS research. I elaborate on each of these below.

Willingness and Ability to Be Immersed

One of the key requirements in ethnographic research is that the researcher enters and becomes immersed in the field for a significant period of time, e.g., a few months to a few years. This implies that the researcher becomes the research instrument, i.e., recording and analyzing data and eventually translating the insights gained for another community. But what does it mean to be a research instrument? What does it take to become immersed?

While physical presence in the field site is a prerequisite of immersion, it is merely the first step in a multi-level process of becoming part of the field. Immersion has a quality of totality and completeness, which implies that physical immersion needs to be accompanied by immersion at the intellectual and emotional level. Allowing oneself to become so completely engulfed in the new locale and culture requires a readiness for learning about oneself and others, and a willingness to give up control. After all, the ethnographic endeavor is about understanding, which means to stand under something. Furthermore, the ethnographic endeavor is intended to let insights emerge from events that occur naturally in the field. This demands sufficient patience, humility and self-doubt to question fundamentally one's own knowledge, perspective and judgment. In my experience, ethnographers need to be willing and able to suspend — though not abandon — their sense of competence, identity and self. Ethnographic fieldwork typically proves to be a transformative experience for the researcher (Coffey, 1999; Kondo, 1990).

My physical immersion was facilitated by the fact that the two field locations were two and three hours' drive away from my home. This created a clear separation between my home life and my life in the field. Physically being *in* the field from Monday morning to Thursday evening made my immersion at the emotional and intellectual level possible. Being able to focus on my role as a learner and research instrument from Monday through Thursday allowed me to develop my identity as an ethnographer without interference from my at-home identity. I did not have to flip-flop constantly between being a competent member of my at-home community and a bumbling learner and misfit in the field.

My experience as a research instrument was dominated by a sense that my self was split into an experiencing self and an observing self. While my experiencing self was engaging in, reacting to, reflecting on and feeling about events in the field, my observing self was monitoring the situation and myself in it. It was as if my observing self was floating above my body, seeing all around me and putting it into words. My observing self thus created a bridge between my experiencing self and my fieldnotes: by transforming my experiences and emotions into language, it became possible for me to capture them into my notebook computer. As I wrote my fieldnotes, my observing self embellished the descriptions of the events 'as they occurred' with my own, immediate interpretations, reactions and emotional responses. This gave my fieldnotes a self-reflexive character, a quality I later drew upon to write up my research in a confessional or reflexive style.

Another factor that aided my willingness and ability to become completely immersed in the field included the fact that I was single. By not having a family or significant relationship to take care of, I was able to devote my entire attention on the fieldwork and my role as a research instrument. I thus had the luxury of being self-absorbed, a state that helped me sustain a high degree of self-reflexivity throughout the project.

It was also helpful that I did not have much prior knowledge of what it was like to be in an American corporation as this translated into feelings of awe, humility and anticipation. I was eager to learn everything about everything; in other words, I was willing to dedicate myself completely to this project in order to understand what was going on. It also meant that I felt more comfortable listening and observing rather than talking and taking action. This orientation was well-suited to participant observation. Furthermore, I noticed lots of details that others more familiar with corporate America might have taken for granted. These included elements of the physical and spatial arrangements within the organization such as office furniture and decorations, and the use of public spaces such as the elevators for the dissemination of information by way of posters and flyers. Also, I was very sensitive to the language used in the corporate environment.

The Availability of a Supportive Thought Community Outside the Field

While I was a research instrument from Monday through Thursday, Fridays were dedicated to meeting my obligations as a Ph.D. student and member of the Management Information and Decision Systems (MIDS) Department at Case Western Reserve University. As Ph.D. students we were expected to attend departmental seminars and meet with invited speakers or job candidates on Fridays. These events did not only help me think about my research in more theoretical terms, but they also helped me maintain my ties with members in the department. Saturdays were usually spent on household chores such as doing the laundry, cleaning, and shopping, but I also spent a lot of the day catching up on my fieldnote writing.

I would spend Sunday mornings collecting my thoughts pertaining to that week's events from the field in preparation for a meeting with my dissertation advisor, Richard (Dick) Boland. Throughout my fieldwork, we met on Sundays for two to three hours to talk about my research, especially the incidents that had happened that week. Typically we would be walking through a local nature reserve during these conversations. I would recount the high points of the week and Dick would ask questions. These conversations helped me identify what information I was using to create narratives and make sense of the field, and they challenged me on my initial interpretations. They also heightened my sensibility to my self in the field and in my reconstruction of it. Through these conversations, we formulated intermediate research questions and strategies for collecting the kind of data that would generate answers to the questions.

These meetings were invaluable for my research because they not only helped me formulate my data collection strategy from week to week, but they also helped me theorize the data while I was collecting it. This is something that many ethnographers fail to do and thereby run the risk of concluding their stay in the field without having the kind of data they need to support the claims and contributions to knowledge they want to make (Kleinman and Copp, 1993).

Another support system presented itself in a summer fellowship program sponsored by the Carnegie Mellon Foundation. This program started just as I completed my fieldwork. Two professors at Case Western Reserve University, one from English and one from History, organized and supervised a semester-long seminar in which Ph.D. students, all of whom were in the throes of writing their dissertations, discussed their work. Participating students were responsible for one seminar session, for which they had to write one chapter of their dissertation. The Ph.D. students that participated were doing research in the humanities. Due to their familiarity with social theories and interpretive research, they proved to be a particularly supportive thought community. Instead of wasting time defending my "soft" research method to them, I could focus on resolving methodological issues with which I was struggling. Also, most probably because they were unfamiliar with information systems and management research, they did not ask the debilitating "so what?" and "where's the contribution?" questions. This meant that I could get on with theorizing the data and writing up the findings.

The Tenacity to Carry On Despite Ambiguity

Due to the social and longitudinal nature of ethnographic research, it is risky especially if it is built around information systems projects, which are notorious for going over time and over budget, for getting postponed, and for being abandoned or even cancelled. Also, not only is it difficult to negotiate initial access and the permission to become a participant observer in an organization, there are no guarantees that this permission will not be revoked during the course of the project.

Negotiating access is an ongoing endeavor, as the permission to observe and ask questions needs to be obtained, either formally or informally, on a continuous, individual basis. Access to information and the right to ask questions is frequently based on the nature of the researcher's relationship with the participants in the field. In order to ask questions, the researcher needs to answer participants' questions. If researchers expect participants to make themselves vulnerable, they have to make themselves vulnerable also. This implies that gaining and maintaining access is emotional work, especially given the potential for personality clashes and disagreements between the researcher and the participants in the field and the very real threat of project cancellation.

I experienced a lot of fear around the issue of access. Not being sure about the scope and direction of my research, especially in light of the fact that the KnowMor project was limping along, was cause for anxiety. The changing direction of my research also raised ethical concerns[10] with respect to identifying what kind of data I had permission to collect and what I was allowed to do with it. Since I relied mostly on informal interactions, participant observation and my personal experience with the work that the systems administrators and to a lesser extent the librarians and the competitive intelligent analysts were doing, there was ambiguity around when I was and when I was not collecting data. For fear of adversely affecting my relationship with the research participants, I did not remind them continuously of my role as researcher, even though I was generally making notes in my notebook thus indicating that I was documenting encounters and conversations. In the end, I relied on the respondent validation process (Silverman, 1993) to resolve any questions about confidentiality and permission to use the material I had collected. Needless to say, this was a very anxious time.

It is also challenging to manage one's own and the host organization's expectations of the research outcomes. Due to the emergent nature of ethnographic insights, it can be difficult to implement the intended research design. This is because, during the course of the research, the ethnographer might gain insights that highlight the implausibility of the intended data collection strategies or the trivial or uninteresting nature of the initial research questions (e.g., Prasad, 1997). Since many organizations agree to give access to the researcher in return for the research results or feedback of a consultative nature, the uncertainty over the final outcome can be quite disconcerting. Coupling concerns of not living up to the expectations of the host organization with uncertainty about what the data means and whether the findings will be deserving of a Ph.D. degree or a publication, can lead to debilitating fear and severe frustration. In addition to managing one's own and the host organization's expectations, the ethnographic researcher must hold firm the belief that he/she will find something interesting. This belief is key to going on.

The Discipline to Write Self-Reflexive Fieldnotes

As I described earlier, my fieldnote writing resulted in a splitting of my self into an experiencing and an observing self. Managing this dual identity was particularly difficult if I was emotionally involved in the incident that I wanted to write up in my fieldnotes. Finding the vocabulary to capture the multitude of images and emotions that my observing self remembered took a lot of time and therefore a lot of discipline. Untangling what had "actually" happened and how I felt about it was particularly difficult. It might have been easier to ignore my personal involvement in composing my fieldnotes, but I do not think that I would have considered my fieldnotes complete if I would not have written long descriptions about the context in which the incident took place and explanations about why I interpreted an event in a certain way.

Fieldnote writing conventions, such as bracketing my own comments, interpretations and explanations not only provided a legitimate space for this contextual material in my notes, but also disciplined my note taking. The template[11] that I used to structure my daily fieldnotes proved to be another disciplining device. Its layout and headings jogged my memory, and the section addressing my research process with separate headings for "People," "Process," "Academic Thoughts," and "Mistakes I Made," encouraged me to contemplate and document my role in the field from multiple perspectives. Also, the section titled "Personal Notes" reminded me to capture explicitly what was going on for me at any given point during the fieldwork.

The Ability to Identify Ways in Which Self-Reflexive Material Makes a Substantive Contribution to IS Research

Even though many advocate the use of reflexive writing to enhance the reliability and criticality of ethnographic research, many of them also caution against the excesses of this mode of representation (e.g.,

Coffey, 1999; Davies, 1999). They highlight that there is a danger that reflexive texts become too autobiographical and "self-serving and superficial, full of unnecessary guilt or excessive bravado" (Behar, 1996, p.14). They can be too personal and self-absorbed and thus result in "'vanity ethnography,' in which only the private muses and demons of the fieldworker are of concern" (Van Maanen, 1988, p. 93). Reviewing the "postmodern" trends in ethnographic writing, Wolf (1992) points out that she is "still ... just a little more interested in the content of the ethnographies we read and write than in the ethnographers' epistemologies" (p. 1).   Behar (1996, p. 18) maintains that self-revelation in ethnographic writing is inappropriately self-serving if the ethnographic subject is vanquished by the autobiographical content:

> To assert that one is a 'white middle-class woman' or a 'black
> gay man' ... is only interesting if one is able to draw deeper
> connections between one's personal experience and the sub-
> ject under study. That does not require a full-length autobiogra-
> phy, but it does require a keen understanding of what aspects of
> the self are the most important filters through which one perceives
> the world and, more particularly, the topic being studied (p. 13).

As I was writing up my dissertation research, I recognized the parallel between the informing practices that I was relying on as an academic knowledge worker, and the informing practices of the knowledge workers I had studied. This insight made it possible for me to use my experience as a knowledge worker and my self-reflexive fieldnotes to generate substantive insights. One of the insights generated through the incorporation of my own experience as a knowledge worker was that informing practices are practical solutions to the inevitable tension between subjectivity and objectivity that knowledge workers face when they create information (Schultze, 2000). It is important to note, that I did not anticipate such an outcome at the start of my research. The idea that my own experiences of creating information as an academic knowledge worker could provide me with a useful mechanism for making sense of the informing practices of the knowledge workers that I was studying, was a consequence of my natural predisposition towards self-reflection and my experimentation with a style of writing that more accurately reflected how the research had been done.

CONCLUDING THOUGHTS

This chapter adds reflexive ethnography to the types of ethnographies that information systems researchers are already familiar with, namely, the holistic, the thick description and the critical ethnography (Myers, 1999). Reflexive ethnography distinguishes itself from these other forms in that it acknowledges the deeply personal and emotional nature of fieldwork, and that it recognizes that textual reconstructions of the field are as much representations of the social group being studied as they are constructions of the researcher's and the academic discipline's identity (Davies, 1999). This makes reflexive ethnographies well suited for cultural critique (Marcus and Fischer, 1986) and criticality (Golden-Biddle and Locke, 1993) because it invites readers to enter the field vicariously, making it possible for them to not only assess the reasonableness of the ethnographer's assumptions and interpretations, but also those of their own culture. By treating personal reactions and emotions as important and legitimate data (Kleinman and Copp, 1993), reflexive ethnography provides readers with a point of entry into the assumptions, practices and taken-for-granted knowledge of the researcher's and their (the readers') own culture.

What revisiting my own experience with doing reflexive ethnography has highlighted is that personal factors such as a predisposition towards reflexivity, family responsibility and the desire to submit oneself to learning, institutional factors such as teaching schedules and tenure clocks, as well as factors associated with the field site such as the proximity to one's home, play an important role in enabling and limiting a researcher's ability to do reflexive ethnography. This is because these factors not only help or hinder the researcher's willingness and ability to get immersed in fieldwork, but they also affect the degree to which the researcher can enter and sustain a state of self-absorption and self-consciousness.

REFERENCES

Agar, M. H. (1986). *Speaking of Ethnography*. Beverly Hills, CA: Sage.

Amit, V. (Ed.). (2000). *Constructing the Field: Ethnographic Field-*

work in the Contemporary World. London: Routledge.

Atkinson, P., and Hammersley, M. (1994). Ethnography and Participant Observation. In N. K. Denzin and Y. S. Lincoln (Eds.), *Handbook of Qualitative Research* (pp. 248-261). Thousand Oaks: Sage.

Barley, S. R. (1988). On Technology, Time and Social Order: Technically Induced Changes in the Temporal Organization of Radiological Work. In F. A. Dubinkas (Ed.), *Making Time: Ethnographies of High-Technology Organizations* (pp. 123-169). Philadelphia: Temple University Press.

Barley, S. R. (1996). Technicians in the Workplace: Ethnographic Evidence for bringing Work into Organizational Studies. *Administrative Science Quarterly, 41*(3), 404-441.

Behar, R. (1996). *The Vulnerable Observer: Anthropology that Breaks your Heart*. Boston, MA: Beacon Press.

Blomberg, J., Giacomi, J., Mosher, A., and Swenton-Wall, P. (1993). Ethnographic Field Methods and their Relation to Design. In D. Schuler and A. Namioka (Eds.), *Participatory Design: Principles and Practices* (pp. 123-155). Hillsdale, NY: Lawrence Erlbaum.

Bourdieu, P., and Wacquant, L. C. D. (1992). *Invitation to Reflexive Sociology*. Chicago: University of Chicago Press.

Brown, J. S., and Duguid, P. (1991). Organizational Learning and Communities-of-Practice: Toward a Unified View of Working, Learning and Innovation. *Organization Science, 2*(1), 40-57.

Bucciarelli, L. L. (1994). *Designing Engineers*. Cambridge, MA: MIT Press.

Button, G., and Harper, R. (1996). The Relevance of 'Work-Practice' for Design. *Computer Supported Cooperative Work, 4*, 263-280.

Carspecken, P. F. (1996). *Critical Ethnography in Educational Research: A Theoretical and Practical Guide*. Routledge: New York.

Clifford, J. (1986). Introduction: Partial Truths. In J. Clifford and G. E. Marcus (Eds.), *Writing Culture* (pp. 1-26). Berkeley, CA: University of California Press.

Clifford, J., and Marcus, G. E. (Eds.). (1986). *Writing Culture*. Berkeley, CA: University of California Press.

Coffey, A. (1999). *The Ethnographic Self: Fieldwork and the Representation of Identity*. London: Sage.

Davies, C. A. (1999). *Reflexive Ethnography: A Guide to Researching Selves and Others*. London: Routledge.

Denzin, N. K. (1997). *Interpretive Ethnography: Ethnographic Practices for the 21st Century*. Thousand Oaks, CA: Sage.

Forsythe, D. E. (1993). Engineering Knowledge: The Construction of Knowledge in Artificial Intelligence. *Social Studies of Science, 23*, 445-477.

Forsythe, D. E. (1995). Using Ethnography in the Design of an Explanation System. *Expert Systems with Applications, 8*(4), 403-417.

Giddens, A. (1987). *Social Theory and Modern Sociology*. Stanford, CA: Stanford University Press.

Golden-Biddle, K., and Locke, K. (1993). Appealing Work: An Investigation of how Ethnographic Texts Convince. *Organization Science, 4*(4), 595-616.

Grudin, J. (1994). Groupware and Social Dynamics: Eight Challenges for Developers. *Communications of the ACM, 37*(1), 93-105.

Hammersley, M. (1992). *What's Wrong with Ethnography? Methodological Explorations*. London: Routledge.

Hammersley, M., and Atkinson, P. (1995). *Ethnography: Principles in Practice*. (Second ed.). London: Routledge.

Kleinman, S., and Copp, M. A. (1993). *Emotions and Fieldwork*. (Vol. 28). Newbury Park: Sage.

Kondo, D. (1990). *Crafting selves : power, gender, and discourses of identity in a Japanese Workplace*. Chicago: University of Chicago Press.

LeCompte, M. D., and Schensul, J. J. (1999). *Designing and Conducting Ethnographic Research*. Walnut Creek, CA: AltaMira Press.

Lyytinen, K. J., and Ngwenyama, O. K. (1992). What does Computer Support for Cooperative Work Mean? A Structurational Analysis of Computer Supported Cooperative Work. *Accounting, Management and Information Technology, 2*(1), 19-37.

Mack, R. (1971). *Planning on Uncertainty: Decision Making in Business and Government Administration*. New York: Wiley-Interscience.

Marcus, G. E., and Fischer, M. M. J. (1986). *Anthropology as Cultural Critique: An Experimental Moment in the Human Sciences*. Chicago: University of Chicago Press.

Miles, M. (1979). Qualitative Data as an Attractive Nuisance: The Problem of Analysis. *Administrative Science Quarterly, 24*, 590-601.

Myers, M. D. (1994). A Disaster for Everyone to see: An Interpretive Analysis of a Failed IS Project. *Accounting, Management and Information Technology, 4*(4), 185-201.

Myers, M. D. (1999). Investigating Information Systems with Ethnographic Research. *Communications of the Association for Information Systems, 2*(23).

Myers, M. D., and Young, L. W. (1997). Hidden Agendas, Power, and Managerial Assumptions in Information Systems Development: An Ethnographic Study. *Information Technology and People, 10*(3), 224-240.

Okely, J., and Callaway, H. (Eds.). (1992). *Anthropology and Autobiography*. London: Routledge.

Orlikowski, W. J. (1991). Integrated Information Environment or Matrix of Control? The Contradictory Implications of Information Technology. *Accounting, Organization and Information Technology, 1*(1), 9-42.

Orlikowski, W. J. (1992). The Duality of Technology: Rethinking the Concept of Technology in Organizations. *Organization Science, 3*(3), 398-427.

Orlikowski, W. J. (1996). Improvising Organizational Transformation Over Time: A Situated Change Perspective. *Information Systems Research, 7*(1), 63-92.

Orlikowski, W. J., and Robey, D. (1991). Information Technology and the Structuring of Organizations. *Information Systems Research, 2*(2), 143-169.

Orr, J. (1990). Sharing Knowledge, Celebrating Identity: Community Memory in a Service Culture. In D. Middleton and D. Edwards (Eds.), *Collective Remembering*. London: Sage.

Orr, J. E. (1996). *Talking About Machines*. Ithaca: ILR Press.

Pentland, B. (1995). Information Systems and Organizational Learning: The Social Epistemology of Organizational Knowledge Systems. *Accounting, Management and Information Technology, 5*(1), 1-21.

Perlow, L. A. (1998). Boundary Control: The Social Ordering of Work and Family Time in a High-tech Corporation. *Administrative Science Quarterly, 43*(2), 328-357.

Perlow, L. A. (1999). The Time Famine: Toward a Sociology of Work Time. *Administrative Science Quarterly, 44*(1), 57-81.

Pickering, A. (1992). From Science as Knowledge to Science as

Practice. In A. Pickering (Ed.), *Science as Practice and Culture* (pp. 1-26). Chicago: The University of Chicago Press.

Prasad, P. (1993). Symbolic Processes in the Implementation of Technological Change: A Symbolic Interactionist Study of Work Computerization. *Academy of Management Journal, 36*, 1400-1429.

Prasad, P. (1997). Systems of Meaning: Ethnography as a Methodology for the Study of Information Technologies. In A. S. Lee, J. Liebenau, and J. DeGross (Eds.), *Information Systems and Qualitative Research* (pp. 101-118). London: Chapman and Hall.

Preston, A. (1986). Interactions and Arrangements in the Process of Informing. *Accounting, Organizations and Society, 11*(6), 521-540.

Reed-Danahay, D. E. (Ed.). (1997). *Auto/Ethnography: Rewriting the Self and the Social*. Oxford: Berg.

Sanday, P. R. (1979). The Ethnographic Paradigm(s). *Administrative Science Quarterly, 24*(4), 527-538.

Schultze, U. (2000). A Confessional Account of an Ethnography about Knowledge Work. *MIS Quarterly, 24*(1), 1-39.

Schultze, U., and Boland, R. J. (2000). Place, space and knowledge work: a study of outsourced computer systems administrators. *Accounting, Management and Information Technologies, 10*(3).

Schultze, U., and Boland, R. J. (forthcoming). Knowledge Management Technology and the Reproduction of Knowledge Work Practices. *Journal of Strategic Information Systems*.

Silverman, D. (1993). *Interpreting Qualitative Data: Methods for Analyzing Talk, Text and Interaction*. London: Sage.

Spitler, V., and Gallivan, M. (1999). The Role of Information Technology in the Learning of Knowledge Work. In O. Ngwenyama, L. D. Introna, M. D. Myers, and J. I. DeGross (Eds.), *New Information Technologies in Organizational Processes: Field Studies and Theoretical Reflections on the Future of Work* (pp. 257-275). Norwell, MA: Kluwer.

Suchman, L. (1993). Technologies of Accountability: Of Lizards and Aeroplanes. In G. Button (Ed.), *Technology in Working Order: Studies of Work, Interaction and Technology* (pp. 113-126). London: Routledge.

Suchman, L. (1995). Making Work Visible. *Communications of the ACM, 38*(9), 56-64.

Trauth, E. (1997). Achieving the Research Goal with Qualitative Methods: Lessons Learned Along the Way. In A. S. Lee, J. Liebenau, and J. I. DeGross (Eds.), *Information Systems and Qualitative Research* (pp. 225-245). London: Chapman and Hall.

Turner, J. A., Bikson, T. K., Lyytinen, K., Mathiassen, L., and Orlikowski, W. (1991). Relevance versus Rigor in Information Systems Research: An Issue of Quality. In H. E. Nissen, H. K. Klein, and R. Hirschheim (Eds.), *Information Systems Research: Contemporary Approaches and Emergent Traditions* (pp. 715-745). North-Holland: Elsevier Science Publishers.

Van Maanen, J. (1988). *Tales of the Field: On Writing Ethnography.* Chicago: University of Chicago Press.

Whyte, W. F. (1996). On the Evolution of *Street Corner Society.* In A. Lareau and J. Shultz (Eds.), *Journeys through Ethnography: Realistic Accounts of Fieldwork* (pp. 9-73). Boulder, Colorado: Westview Press.

Wolf, M. (1992). *A Thrice-Told Tale: Feminism, Postmodernism, and Ethnographic Responsibility.* Stanford, CA: Stanford University Press.

Wynn, E. (1991). Taking Practice Seriously. In J. Greenbaum and M. Kyng (Eds.), *Design at Work* (pp. 45-64). New Jersey: Lawrence Erlbaum.

Zuboff, S. (1988). *In the Age of the Smart Machine.* New York: Basic Books.

ENDNOTES

1   It is important to note that there is considerable debate among anthropologists regarding the legitimacy of field sites. Whereas only exotic, faraway places were typically considered 'true' field sites, more and more ethnographers are engaging in anthropological inquiries 'at home' where their informants are their family, friends and neighbors (Amit, 2000).

2   Taking a critical perspective of fieldnotes, Kleinman and Cobb (1993) suggest that fieldnote writing gives ethnographers a sense of identity and a reason for their presence in the field through hours, and even days, of boredom, uncertainty and frustration when apparently nothing worthwhile is happening. The discipline of fieldnote writing can also be seen as a crutch that provides the fieldworker a sense of stability and purpose during periods of great uncertainty.

3   While I was already immersed in my Ph.D. fieldwork, a Ph.D. student from another university told me that she decided not to do intensive, qualitative research because she had been told that she 'would never get a job.'

4   Even though the meanings of hard and soft information are not well established and the distinction between the two concepts is problematic, there is a general perception that hard information (e.g., quarterly earnings) is more factual than soft information (e.g., a rumor). Furthermore, hard information is typically arrived at through formal procedures and represented in highly structured and frequently quantitative formats, whereas soft information is generated in more informal, narrative ways and frequently represented as text.

5   My use of the terms 'subjective' and 'objective' need to be understood in the context of the social constructionist orientation, which posits that neither a position of absolute subjectivity nor of absolute objectivity is possible. Instead, both become relative positions in the intersubjective social consciousness. Thus the scientistic *subject-object dualism*, which posits that reality is made up of objects that stand in fundamental opposition to subjects, i.e., individuals with minds and feelings, is replaced by a *subject-object duality*. This represents a view of subjects and objects as inextricably intertwined and mutually constitutive.

6   The power that accrues to the author.

7   A pseudonym.

8   To assure confidentiality, the line of products that US Company manufactured is omitted intentionally.

9   All participants' names used in this text are pseudonyms.

10   The ethical issues related to ethnographic research are considerable, even when the research is covert. There are numerous texts that address them (e.g., Amit, 2000; Okely and Callaway, 1992; Davies, 1999). I found Hammersley and Atkinson's (1995) discussion the most comprehensive and useful.

11   Before entering the field, and based on my readings of the ethnographic method (e.g., Silverman, 1993), I developed template documents in my word processor into which I could capture my fieldnotes. The template I used the most was the "DAY" template (for an exact copy, please refer to Schultze,

2000). It contained sections into which I captured the events of the day. The section headings were "Date," "Location," "Project Phase," "To Do's," "Main Events," "Small/Odd Events," "What I Learned," "Plans for Tomorrow," "Calendar of Today's Events," "Detailed Description of Main Event," "About my Research: People," "About my Research: Process," "About my Research: Academic Thoughts," "About my Research: Mistakes I Made." and "Personal Notes."

Chapter V

An Encounter with Grounded Theory: Tackling the Practical and Philosophical Issues

Cathy Urquhart
University of the Sunshine Coast, Australia

INTRODUCTION

The purpose of this chapter is to explore the practical and philosophical issues of applying the grounded theory approach to qualitative research in Information Systems. Over the past decade, we have seen a substantial increase in qualitative research in general (Klein, Nissen and Hirschheim, 1991; Walsham, 1995; Markus, 1997; Myers, 1997; Myers and Walsham, 1998; Klein and Myers, 1999; Walsham and Sahay, 1999; Trauth and Jessup, 2000; Schultze, 2000) and also an increase in the use of grounded theory (Toraskar, 1991, Orlikowski, 1993, Urquhart, 1997, 1998, 1999a, 1999b; Adams and Sasse, 1999, Baskerville and Pries-Heje, 1999, Trauth, 2000). Over the past three years, the most frequent request I have had from postgraduates is for some insight into the 'how-to' of coding and grounded theory.

Obviously these observations are not unconnected, as an increase in the use of qualitative methods in information systems results in a search for interesting, meaningful and useful ways to analyse data. I believe grounded theory technique does provide precisely that—an interesting, meaningful and useful way to analyse data—but there are also a number of practical and philosophical issues associated with its application, which I propose to explore in this chapter.

My own experience of using grounded theory in practice raised a number of accompanying philosophical issues along the way, and it is only when we attempt to use grounded theory in practice that the philosophical issues become apparent. By philosophical issues I mean issues of ontology[1] and epistemology[2], and how the method itself might imply or confer a certain position of either interpretivism or positivism.

However, as grounded theory is primarily a method of data analysis, it is necessary to grapple with the 'how-to' issues first before understanding how the method in use might have ramifications for a particular research philosophy or approach. This chapter therefore addresses itself to two major aspects.

Firstly, there is a shortage of literature in the form of practical guidance on the 'how-to' of grounded theory technique, and this chapter proposes to remedy this by offering a detailed example of application. For instance, one issue with grounded theory technique that researchers seem to encounter is the difficulty of scaling up their analysis to larger themes. This is not surprising when one considers that, as a coding method, it is essentially a *bottom-up* technique in relation to the data, and begins at the word or sentence level.

It might well be that grounded theory technique has gained popularity in our field precisely because it does offer relatively well signposted procedures for data analysis and it is well known as a method. Those procedures *seem* well signposted—until the researchers end up wrestling with such practical issues as whether to see the data item in front of them as a property, dimension or category, or whether the core category or categories they have selected is the right one, or how to write up their analyses. Obviously each researcher has to find his or her own way with regard to their particular analysis, but clearer explications of coding methodologies in published work and specialised seminars are needed in our discipline if we are to get the best out of what grounded theory and other coding approaches can offer.

There are other approaches to coding, such as using predetermined codes or taking a 'middle order' approach (Dey, 1993), where some preliminary distinctions in the data are made. Thus it is important to realise that grounded theory technique is but one way to 'code' data and it needs to be seen in the context of other approaches to analysing and coding data. I can do no more in this chapter than direct readers to Dey's (1993) excellent material on coding qualitative data, and also to Miles and Huberman's (1994) chapter on coding and analysing data, which provide a wider perspective on coding qualitative data.

Secondly, there is the issue of how we in the IS research community 'adapt' and use grounded theory in our field. Grounded theory does come with its own philosophical baggage (Annells, 1996) and so comes the question of whether it is being used primarily as a technique for analysing data, or as a research philosophy in its own right. As most IS studies using grounded theory technique tend not to be 'pure' studies, it is most likely the former. That said, there still are a number of misunderstandings about the use of grounded theory technique, associated with the IS field's tendency to import selectively from other disciplines, without fully understanding the genesis of a given theory or technique. This tends to be manifested by quoting only one reference on the theory or technique, rather than examining it more deeply, as this chapter intends to do.

A DEFINITION OF GROUNDED THEORY

Grounded theory had its origins in the 1967 book, *The Discovery of Grounded Theory*, by its co-originators, Barney Glaser and Anselm Strauss. They developed a research methodology that aimed to systematically derive theories of human behaviour from empirical data, and this could also be seen to be part of the symbolic interactionist (Blumer, 1969, in Kendall, 1999) reaction to 'armchair' functionalist theories in sociology (Kendall, 1999, Dey, 1999). This methodology was also a contribution to the field of symbolic interactionism in that Blumer, (1969, in Kendall, 1999) and the famous Chicago School of Sociology had long maintained that there was a need for a special methodology for the study of human behaviour (Kendall, 1999).

From its inception in 1967, grounded theory spread fairly quickly as an accepted qualitative research method, particularly in the health field. There were several more publications elaborating on, developing, and later, debating the method (Glaser and Strauss, 1967; Glaser, 1978; Strauss, 1987; Strauss and Corbin, 1989, 1990; Glaser, 1992, Strauss and Corbin, 1994; Glaser, 1995, 1998). Over the years, grounded theory method developed into two distinct variants, one favoured by Glaser, the other by Strauss (Melia 1996). A very public disagreement between the co-originators on the publication of Strauss and Corbin's book in 1990 compounded those differences. This controversy, and its possible consequences for those using the method, will be examined in detail later in the chapter. For now, I would like to proceed with a fairly unproblematic definition of grounded theory and illustrate the issues that occurred in practice by way of an example (including how the controversy was dealt with in that research).

Dey (1999), drawing on Creswell (1998), gives the following very workable definition of grounded theory:

1. The aim of grounded theory is to generate or discover a theory.
2. *The researcher has to set aside theoretical ideas* to allow a 'substantive' theory to emerge.
3. Theory focuses on how individuals interact in relation to the phenomenon under study.
4. Theory asserts a plausible relation between concepts and sets of concepts.
5. Theory is derived from data acquired through fieldwork interviews, observations and documents.
6. Data analysis is systematic and begins as soon as data is available.
7. Data analysis proceeds through identifying categories and connecting them.
8. Further data collection (or sampling) is based on emerging concepts.
9. These concepts are developed through *constant comparison* with additional data.
10. Data collection can stop when new conceptualisations emerge.
11. Data analysis proceeds from 'open' coding (identifying categories, properties and dimensions) through 'axial' coding (examining conditions, strategies and consequences) to 'selective' coding around an emerging storyline.
12. The resulting theory can be reported in a narrative framework or as a set of propositions (Dey, 1999, pp 1-2).

The phrases in italics, viz., *'the researcher has to set aside theoretical ideas'* and the idea of concepts being developed through *'constant comparison'*, are two key features of grounded theory that make adopting this approach very different from any other approach to analysis of qualitative data.

The idea of *setting aside theoretical ideas* implies that the researcher does not look at existing literature. This is not in fact an accurate representation of grounded theory—the position of both Glaser and Strauss on this issue is far more subtle. The injunction about literature seems mainly designed to ensure that the researcher takes an inductive rather than deductive approach, and listens to the data rather than imposing preconceived ideas on the data. However, advisors or supervisors, and perhaps also colleagues, when confronted with a researcher wanting to use grounded theory, may assume that the researcher is proposing to use a non rigorous method. Nothing could be further from the truth as we shall see in succeeding sections.

The idea of *constant comparison* is at the heart of grounded theory as a method, and can be seen as nothing more than an enlightening rule of thumb which assists researchers to understand the process of analysis. Put simply, constant comparison is the process of constantly comparing instances of data that you have labelled as a particular category with other instances of data, to see if these categories fit and are workable. If they do, and the instances mount up, then we have what Strauss (1987) and Glaser (1992) call 'theoretical saturation'.

PRACTICAL ISSUES—AN EXAMPLE OF A GROUNDED THEORY STUDY

This section tells the story of how I first came to use grounded theory, and is to some extent a typical tale of a researcher using grounded theory for the first time. For instance, I was not acquainted with all the works of Glaser and Strauss, and it took some time for me to realise that some advice on grounded theory was contradictory. Also in common with most people using grounded theory for the first time, I had a great deal of difficulty with the analysis—but was richly rewarded in terms of concepts yielded. The story begins in 1995 when I was looking for an approach to investigate analyst-client interaction, and is told below.

The Research Problem

From my experience as a systems analyst, I was convinced that communication between analysts and clients was a vital building block in the systems lifecycle. I was curious about the process of communication and how concepts were built up in a conversation between analyst and client. So, I started out with a very general research problem: *How do analysts and clients approach early requirements gathering?*

It should be noted that the research *questions* came much later, and were a product of the analysis. Dey (1993) notes that this is often a natural consequence of qualitative analysis. Interestingly, Glaser (1992) says much the same:

"out of open coding.. theoretical sampling and analyzing by constant comparison emerge a focus for the research" (Glaser, 1992, p. 25).

It wasn't until much later of course, that I discovered that it was a reasonable thing to do, to have only a research problem rather than successive questions. At the time, well-meaning advisors would ask me, 'What are your research questions?' as a way to get me to focus. This was difficult —given there wasn't much literature in the field, so I couldn't get my research questions from there. I was also puzzled as to why my experience as an analyst seemed irrelevant and was termed 'anecdotal' by many people. Yet it was this very experience that gave me a passion for researching the phenomenon of analyst-client communication. Having worked as a systems analyst, I wanted to do some research that would contribute to understanding the process, and also feed back some of these ideas into practice.

The Decision to Use Grounded Theory

It was with relief that I turned to grounded theory, as I felt it gave me justification for going straight into the field, which I intuitively wished to do. In a phrase, it got me 'off the hook' – I no longer had to look at literature which I felt was irrelevant to the phenomenon, and construct hypotheses which might or might not occur.

Retrospectively, it is easy to see why grounded theory was so suitable – I wished to study a process (analyst-client communication) and there was very little theory available, so of course an inductive approach was appropriate. Like most people coming to grounded theory for the first time, I read the Strauss and Corbin (1990) book. The thing that stood out in that book for me was that it did not dismiss my previous experience and confirmed that yes, this could be a source of a research problem/interest.

That said, I did look at other methods of analysing conversational data and found them inappropriate for the research problem. In the field of information systems, the only major studies of analyst-client interaction I located (Guinan, 1988; Tan, 1989) used content analysis and independent measures to find relationships among variables. Most approaches to analysing texts seemed to fall either into structural or processual aspects, even though they cannot be easily abstracted from each other (Candlin, 1985), and neither approach would allow me to track the whole process while also looking at how meanings might evolve and change. Also, Strauss's (1987) suggestion that the researcher open code (allocate initial codes) line by line was entirely germane to the anticipated unit of analysis of the study.

Case Study Design

The six case studies that comprised the study on client-analyst interaction were all carried out in public sector agencies in Tasmania, Australia, where I was living at the time. These public agencies tended to represent larger concentrations of information system development activity in the state of Tasmania. I asked IS managers if systems analysts in their employ were undertaking development work, and would be willing to participate in the research project. The criteria for inclusion in the project were that the development work was at an initial stage (generally the first or second meeting between analyst and client) and that the interaction should be about the development of a new system or substantial amendment of an existing system.

The research strategy adopted for the case studies was essentially to examine an analyst client interaction indepth, and to collect as much contextual information around that interaction as possible. This echoes Pettigrew's (1985) contextualist approach, and also incorporates a self reflexive element (Schön 1983), where participants individually and together reflect on the interaction. From my philosophical position as a constructivist, the data sources surrounding the interaction enabled me to find varying social constructions on the interaction, which were subsequently distilled into a consensus construction using hermeneutic techniques (Guba and Lincoln 1994). The case studies employed multiple data sources in order to gain those different viewpoints or constructions, with a videotaped interaction between analyst and client as the centrepiece of the case study. Table 1 shows the data sources and the order of collection.

Open Coding – The First Step
of Grounded Theory Analysis

Having successfully collected the data, I commenced open coding the first transcript of the first case. This involved looking at each line, allocating codes to words or groups of words, and was an extremely laborious process. For a 10-page transcript, the coding effort was approximately 60 hours, but this was nothing compared to the mental effort. I felt lost and uncertain about the process, but persisted to the end. For a period of two months, the transcript followed me everywhere -

Table 1: Data Sources in the Case Studies

Data Source	Order of Collection
Paragraph on issues to be discussed during the interaction	Submitted approximately 10 days before the interaction by each participant
Individual interview	Audiotaped on the day of the interaction, prior to the video-taped interaction
Individual questionnaire on background and training	Administered prior to the videotaped interaction
Interaction between analyst and client	Videotaped after both participants completed interview and questionnaire
Review of interaction by analyst and client	Videotaped after the interaction
Individual interview	Audiotaped after review

including on top of a mountain in Victoria's Grampians while on holiday.

It would have been at this stage that a real-life coding seminar would have been most useful—Strauss's (1987) book gives a transcript of a coding seminar where a researcher, Strauss and other graduate students discuss the coding of a given bit of data. Instead my long-suffering partner was kind enough to debate these meanings and codes with me at this stage, as I encountered the problems that most people do when first using grounded theory. First, it was difficult to know whether the codes I was assigning would be interpreted the same way by other people. I can see now that this was not a problem from an interpretive perspective, but nevertheless there was an anxiety as to whether my interpretations were at all plausible. Secondly, it was hard to not to be influenced by all that I had read about conversation, and not to apply concepts from elsewhere. Thirdly, it was hard to decide at what level to 'chunk' the data – words or sentence level, Strauss's (1987) advice notwithstanding.

I found Dey's (1993) excellent material on coding and categorisation extremely helpful at this point. Dey (1993) outlines three aspects to qualitative data analysis; describing, classifying and connecting. "Describing" is simply providing context by providing a comprehensive description of the phenomena under study. Klein and Myers (1999) advocate much the same thing in their principle of contextualisation, which of course means that the reader is provided with enough context to be able to assess your analysis. "Classification" and "connection" are at the heart of coding. Classifying occurs by allocating an analytic category or concept to a data 'bit', and connecting by thinking about the relationship between categories. It is the relationship between the categories that can be seen as the main engine of theory building.

Figure 1 shows a fragment from the transcript of Case 1 (lines 5-15). Table 2 provides details of the very first open coding of this fragment, carried out at the word and sentence level, together with some comments added at the time. This transcript dealt with a conversation between an analyst and a client about a system which supported the Tasmanian Student Assistance system and which required amendment. In the first part of this conversation, the analyst seems to be trying to gain ownership (indicated by frequent use of 'we'), set an agenda, and also discover some aspects of the scope of the existing system.

Eventually, on a later pass of the transcript I hit on the idea of using *topics* as a unit of analysis. This became both a convenient way of chunking the data and a way of understanding the transcript as a whole.

Figure 1: Fragment from Transcript

5 Analyst.	Umm and basically what I've sort of got down here is the data base is about keeping statistics of approved and non-approved applicants, or students, for a Student Assistance Scheme.
6 Client.	Mmm.
7 Analyst.	Basically we are looking at umm basically how the data base works and possibly some of the points that we are looking, particularly about improving.
8 Client.	Mmm, mmm.
9 Analyst.	You know recording of information.
10 Client.	We do it by school and ..
11 Analyst.	Yeah by school, and the system of applicants and enquiries from public usability recording, just general things that we are thinking about as we are going along.
12 Client.	Yes. Mmm, mmm.
13 Analyst.	But to get that sort of point what I've got is sort of we need to try and work out, or I need to work out what the actual data base does and how it functions at the moment?
14 Client.	Mmm, mmm. Yes, yes.
15 Analyst.	But we'll be able to look at umm what sort of changes we can make to improve things?

I can now see this as an example of theoretical sensitivity (discussed later in the chapter) where what I had read actually helped to build the theory without compromising its emergence. Details of some of the topics for Case 1 (there were 44 in all) and their accompanying open codes are given in Table 3.

Table 2: Early Open Codes for First Transcript

Line No	Early Open Codes
5	assertion ('basically') deprecatory, document ref, system ref ('about'), system function, clerical procedure name * Comment: analyst introduces what he thinks the system does and what it supports. Use of word 'basically' Note from his point of view, the IT—database, precedes in conversation its user function - the Student Assistance Scheme.
6	Agreement (confirmed by a nod)
7	assertion ('basically'), 'we', assertion ('basically'), system ref, discussion objectives (how it works and points we are thinking about improving), 'we'
8	Prompt
9	'you', detailed objective * so recording information is one of the points thought about improving
10	objective scope check (by school?)
11	confirmation, detailed objective, we, catch all objective 'just things we are thinking about as we are going along' * Comment: hint here that other 'things' will be considered during conversation 'as we are going along'
12	Agreement
13	'I-you' (pronoun shift), analyst-objective
14	Prompt
15	'we', discussion objective * Comment: discussion objective—factual (i.e., let's look at how the database works, and subjective (what sort of changes can we make to improve things?)

In Table 3, one can see how chunking the data into topics has enabled the coding—at the same time, it can be seen that the initial open codes did yield some important ideas which eventually became key findings, notably the analyst's use of *agenda setting* and also his method of *scoping*, both of which were apparent in the early open coding.

Dey (1993) also helped me realise an important point—coding is done at an analytic level, not a descriptive one. The difference between the codes presented in Table 2 (early open codes) and Table 3 (codes within topics) are instructive—I moved from simply labelling descriptively to labelling analytically. Interestingly, the comments made in Table 1 give a hint of where I was to end up for that particular fragment

Table 3. Open Codes Allocated to Topics

Topic No	Topic Name	Open Codes
1, 4	Our last meeting	*agenda setting, conversation topic*
2	Points of possible system improvement	*conversation topic*
3	How to discuss possible improvements	*conversation topic, scoping*
5, 32	Role of database in assessment	*scoping (many references)*
6	Need for process to help with assessment	*scoping, problem identification*
7,12	Information received from schools and relationship to database	*scoping, information type*
8	Role of schools with regard to assessment	*scoping, process identification*
9, 11	Role of schools with regard to scheme	*scoping, process identification, information type*
10	Doing one's tax return	*rapport building, personal disclosure*
13	Summary information from schools	*forward reframe, information identification*
14	Process of assessment	*process identification*
15, 20, 22, 24, 26	Information input to database	*information identification (many references), process identification, key searching*
16, 18, 21, 33, 35, 37	Links from information input to applicant	*key searching (many references), information identification, process identification, scoping, imagining, metaphors*

- with a category of *agenda setting*. In the early coding, I could see the analyst was attempting to structure the conversation but did not know how to name it. Also, it should be noted that I could only be sure about *agenda setting* once I had coded instances in the other cases, using the principle of constant comparison. This account of the coding of only a fragment of the transcript—should also give a flavour of how iterative the process was, and indeed further iterations occurred all the way along.

Thus, the 60 long arduous hours spent in open coding was richly rewarded, as it provided the basis for a substantive theory about how analysts and clients go about determining requirements. Open coding of successive transcripts was much quicker, as core categories emerged and the theory 'filled in' as it were.

Axial Coding

Having successfully 'open coded' the data, I commenced axial coding – the stage where categories and relationships between categories are supposed to emerge. It is also at this stage that the open codes are grouped into categories and subcategories, and indeed some open codes become categories in their own right.

It was while using the coding paradigm suggested by Strauss (1987) which asks the researcher to examine the data for *conditions, interaction among the actors, strategies and tactics*, and *consequences*, that I came across my first difficulty. Put simply, I found it difficult to apply the coding paradigm, and the relationships between codes and categories hard to discover. The relationships between codes/categories/concepts (whatever you have decided to call your labelling of data bits) is critical in generating theory. What I did instead was to use the coding paradigm to give me some ideas, to help me draw some preliminary distinctions in the data. These initial distinctions, I hoped, equated to initial core categories. The relationship between these initial core categories, the coding paradigm items, and some initial open codes is represented in Table 4.

At this point then, I started to part company with 'pure' grounded theory or at least the version espoused by Strauss and Corbin (1990). I needed another way to think about the relationships between my initial core categories as I had broken apart Strauss and Corbin's (1990) paradigm to make it useful to me. Had I known about the 18 theoretical coding families proposed by Glaser (1978) at the time, I might well have used them, as these represented alternative coding paradigms.

Like a number of researchers before me, I put my faith in the Strauss (1987) and Strauss and Corbin (1990) works, as I had no reason to think that they did not represent all there was to know about grounded

Table 4. Using the Coding Paradigm to Draw Distinctions in the Data

SAMPLES OF INITIAL OPEN CODES	CODING PARADIGM ITEMS	INITIAL CORE CATEGORY
acting out, imagining, vivid description, posited action, prop, reframe	Interaction among the actors Strategies and tactics	INTERACTION
information source, information type, document ref, computer system ref, clerical system ref, information link, process identification, condition, client action	Conditions Consequences	CONCEPTUALISATION

theory, and there was no reference to theoretical codes in those works. So, I searched for another way of thinking about relationships and used Spradley's (1979) domain analysis which gives useful relationships like *'is a kind of'* and *'is a way to'*. Later, I found Glaser's (1978) coding families, particularly the 'strategy' family, useful prompts not only when considering relationships but also when selectively coding (that is, coding around a core category).

It is also instructive to note that, at this stage, I had no knowledge of the difference between Glaser and Strauss that had occurred in 1990, and that a large part of it centred round the 'forcing' of data into a coding paradigm. Certainly, that particular coding paradigm did not work well for the research problem I was engaged in. Kendall (1999) describes a similar experience where she describes spending two years working and reworking the paradigm model proffered by Strauss and Corbin (1990). At that time, she too had little appreciation of the criticism of the Strauss and Corbin paradigm. The details of the dispute will be spelled out in detail in a later section, but I will remark at this point that when I fell upon Glaser's (1978, 1992) works, it was with something like relief, as his remarks tallied very much with my experience of using grounded theory. Since then, I have come across researchers who have successfully used the Strauss and Corbin coding paradigm. It is interesting to speculate that there might be two problems with the coding paradigm. Firstly, whether it should be applied at all, as Glaser (1992) contends, and secondly, whether the paradigm itself might not be suitable for studying certain phenomena such as conversation. At the same time, I have Strauss and Corbin (1990) to thank for making grounded theory accessible to a novice, though using the coding paradigm was for me a difficult experience. For now though, let us return to the coding story.

At this stage, I had two emergent core categories: one pertaining to Interaction, the other to Conceptualisation. Once I started to selectively code for these, I realised I had a problem—one seemed buried in the other, as is the nature of speech. The next section explains how the core categories were reorientated as a result of reviewing the axial coding results.

Reviewing Axial Coding and Proceeding to Selective Coding.

When I attempted selective coding (coding around the core categories) I could see that the codes did not always fit neatly into either category, indicating that these were not 'true' core categories. For instance, *agenda setting*, where an individual laid out the topic and associated issues, could be seen as pertaining to both core categories of Interaction and Conceptualisation. After a great deal of discussion, and agonised thought, I reorientated the core categories to make them more true to the codes from whence they came.

Figure 2 illustrates the problem (left-hand box), and the accompanying resolution (right-hand box), resulting in two 'new' and 'core' categories, Conversational Strategy and Conceptual Strategy.

Glaser (1978) signposts a core category as being a dimension of the research problem, and indicates that it can also be a process. Given the processual nature of the research problem – *how* do analysts and clients reach shared understanding – this would not seem to be an unreasonable proposition. Given also the previous use of the coding paradigm to focus on strategies and tactics used by analyst and client, I realised that this might be one of the dimensions of the problem.

Returning to the *how* aspect of the research problem, if the category of Conceptualisation was characterised in an activity central to requirements gathering, then this would overcome the difficulty connected with its degree of abstraction and facilitate analysis. It could be related

Figure 2: Reviewing the Core Categories

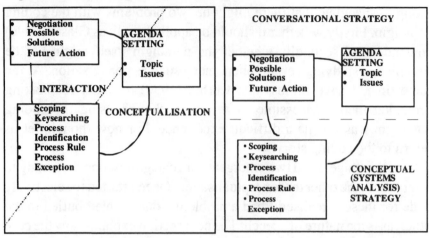

more firmly to the process of early requirements gathering by its renaming to Systems Analysis Strategy.

So, I reformulated the categories as one core category which now represented a process – *strategies in early requirements gathering* with two sub categories – *conversational strategies*, and *systems analysis strategies*. The reformulation of the core categories are illustrated in Figure 3.

Thus the research problem – how do analysts and clients approach early requirements gathering? – revealed its first research question or dimension as expressed by the core category of Strategies in Early Requirements Gathering. This translated into the research question 'what strategies and tactics do analysts and clients employ during the process of early requirements?' Dey (1993) remarks that questions vaguely formed at the outset may be considerably redefined and reformulated by the time the final stage of analysis is reached. Certainly we can see here how the successive reanalysis of the codes and categories aided in the definition of the first research question.

Formulating the categories in this manner allowed that some of the tactics previously identified could be used in a number of circumstances. For instance *metaphors* are a device used to aid understanding in a variety of conversational situations. In Case 1, *metaphors* are clearly a tactic, as they occur both in *imagining*, and *reframing* and many other instances.

Table 5 gives examples of how the codes previously identified, together with lower level codes identified in the axial coding phase,

Figure 3: Reorganisation Into One Core Category

were reclassified as strategies and tactics. They were then related to the new subcategories of conversational strategies and systems analysis strategies. The lower level codes, designated tactics, could be drawn upon by a range of strategies in either subcategory. Some of the individual codes were also reformulated at this point; for instance, *process rule* was deemed to be a tactic rather than a strategy, and seen as supporting the strategy of *process identification*. Similarly, *future action* was recast as a tactic supporting the strategy of *negotiation*. The strategy of *possible solutions* was discarded as it was seen to be encompassed in the tactic of *future action*.

Table 5: Reclassification of Codes into New Categories, Strategies and Tactics

Category	Strategies	Sample Tactics
CONVERSATIONAL STRATEGIES	Negotiation	*posits, future action, forward reframe, problem identification*
	Agenda setting	*conversation topic, issues*
	Rapport building	*'we', joint ownership, personal disclosures*
SYSTEMS ANALYSIS STRATEGIES	Key searching	*posits*
	Information identification	*information type, exemplification*
	Process identification	*posits, process rule, process exception, problem identification*
	Scoping	*posits, information typing*
	Imagining	*metaphors, vivid description, dialoguing, exemplification*
	Reframing	*metaphors, forward reframe*

Table 5 illustrates one of the major issues with coding, in that it is only later in the process that one can decide if a given category is in fact a property of another. For instance, *metaphors* (as a tactic) could be seen as a property of *imagining* and also *reframing,* which are both subcategories. In my experience, one of the stumbling blocks as one transits from open coding to axial coding is deciding which open codes are actually properties of another open code. This in turn may be the basis of an identification of a new category or subcategory. Some codes get elevated, some get downgraded, and the decisions that underly this are founded in *constant comparison*, where patterns in the data become obvious. Glaser (1992) described the procedures of constant comparison in this way:

> The first pertains to the making of constant comparisons of incident to incident, and then, when concepts emerge, incident to concept, *which is how properties of categories are generated* (italics added). The second is asking the neutral question..what category or property does this incident indicate? (Glaser, 1992, p. 39).

This process became easier as I became more practiced at it with the first transcript, and came into its own when examining the transcripts of successive case studies. These transcripts were subjected to *selective* coding where I was only coding for the core category. In this way the core category became *saturated,* that is fully rounded out by reference to the data, and resulted in a substantive theory of the strategies and tactics analysts and clients used in early requirements gathering.

Using Theoretical Memos and Integrative Diagrams

A vital plank of the developing theory was the use of theoretical memos (sometimes called analytic memos) and integrative diagrams.

The writing of theoretical memos is strongly recommended by Strauss (1987), Strauss and Corbin (1990), and Glaser (1978, 1992) as part of the process of developing grounded theory. The idea is that, whenever a researcher is struck by an idea during coding, they should break off at that point and write a memo to develop the ideas. Their role is best expressed by Glaser (1978).

> Memos are the theorizing write-up of ideas about codes and relationships as they strike the analyst while coding (Glaser, 1978).

The importance of this activity cannot be overemphasised—while writing memos on the possible relationships between codes, one is developing possible theory. Memos can be seen as a private diarising that the researcher engages in, where the researcher is also freed from constraints such as the need to write in a polished fashion. Memos also enable the researcher to concentrate on creatively generating ideas about the data.

I found the use of theoretical memos invaluable, to the extent that modified versions of them found their way into various papers, indicating that the ideas developed in them had gone some way to explaining the emergent theory, both to myself and others.

Strauss (1987) recommends the use of integrative diagrams, as a way of integrating threads of the emergent theory. Drawing diagrams for the study was invaluable from two perspectives. Firstly, it was a way of visualising relationships between categories. Secondly, it was a way of breaking out of the necessarily linear nature of written theoretical memos. When explaining the ideas to others, the diagrams proved to be a useful communicative tool. More importantly (and this applies to memos too) they provided a different vantage point from which to view the developing theory, which was helpful during coding periods where conceptualisation seemed to plateau.

As such, both integrative diagrams and theoretical/analytic memos of course are helpful as a general tool for researchers, not just those following the grounded theory approach. Examples of both can be found in the appendix to this chapter.

Dealing with the Literature

The grounded theory approach to dealing with relevant literature has already been alluded to, and will be discussed in detail in a later section. In the study described here, the bulk of the literature searching was carried out *after* the substantive theory was developed. That said, a preliminary theoretical framework, which encompassed four broad aspects of analyst-communication, was used to guide the case study design, and later, literature was organised under these four headings.

The framework ensured some *theoretical sensitivity* (see section "What is Theoretical Sensitivity..."). Even though I was drawing on a range of literature both in information systems and outside it, my thinking was not overly coloured by literature when coding, for two reasons:

First, I delayed the bulk of the literature searching until the substantive theory was developed, enabling me to look at *relevant* literature. Secondly, and this is common when studying some phenomena in information systems, the possible applicable literature spanned a wide range of fields outside information systems, including social cognition, semiotics, speech act theory, and communication rules to name but a few.

Thus delaying some of the literature searching – and not committing to a view of the world proffered by another discipline – made sense in the context of the phenomena being studied. Once various themes around the core category were settled upon, it was straightforward to relate these to the literature.

Strauss (1987) states that contradictory analyses from relevant literature need to be grappled with, and I found this most useful advice. The process of relating the substantive theory with the literature was by turns reassuring and challenging—reassuring when I could see my developing theory embellished, strengthened and corroborated by the literature, and challenging when the literature seemed to contradict the substantive theory. When the literature seemed to contradict the substantive theory, I went back to the data from whence that particular element had come, and reconsidered the analysis. This process can be seen to resonate with two of Klein and Myers' (1999) principles for conducting and evaluating interpretive field studies, namely the principles of abstraction and generalisation, and of multiple interpretations.

Writing Up the Study

When writing up this study, I opted for a narrative framework, where the codes were part of the explanation of the dialogue. I provided a vignette[3] of each case, then proceeded to use various bits of dialogue to illustrate various concepts. One of the advantages of grounded theory is its close tie to the data (Orlikowski, 1993), and I think this confers a further advantage to the researcher—confidence in the findings and the theory put forward. It was easy to point at various parts of the data as illustrating certain concepts, as they had been coded that way previously.

For instance, I could give examples of each strategy and tactic identified in the transcripts. Figure 4 provides a fragment that illustrates this style of write up (Urquhart, 1999b). Interested readers are referred to Urquhart (1997, 1998a, 1999a, 1999b) for further examples.

What Did the Grounded Theory Approach Give to the Study That Other Approaches Could Not?

An important question is whether the application of a grounded

Figure 4: A Fragment of the Write Up for Case 1

The analyst asks:

"OK, so when you put in the summary information you put in, you put in the number, ..does each number..apply to each application?"

He is *keysearching*—looking for links in the data that could be subsequently used to access that data. The client replies:

"yes it does"

The analyst proceeds to clarify the precise nature of the link:.

"so you sort of have another code number or something for each applicant that gets put into the database?"

The client realises where his conclusion might be heading and says:

"It's not, it's not a reference to the student, at the moment it can't be referenced to any individual student.."

She uses this temporary termination of this line of questioning to proceed with some *agenda setting* of her own, via some *problem identification*. She says:

"but we don't have any student records there, so..the capacity you know..twenty seven thousand or so. SACS was going to solve this problem."

---

theory approach resulted in a better study than one using more conventional methods. My answer in this case would have to be a resounding yes, though this is of necessity a subjective view.

Let us consider some of the characteristics of this study and *why* grounded theory was used. Firstly, the phenomena under investigation, analyst-client communication, had only two detailed investigations in information systems literature (Guinan, 1998; Tan, 1989), both of a quantitative nature. Therefore it was easy to follow the injunction of grounded theory not to be overly influenced by literature in the field, as there was not a body of work to follow. Secondly, the paucity of the literature also implied that exploratory, descriptive research requiring an inductive approach would be appropriate. Thirdly, the research problem of interest was the process of communication, and again grounded theory is particularly suited to the investigation of processes (Glaser, 1978). Finally, the grounded theory approach provides a wealth of literature on coding and analysing data, though it should be noted that the original work on grounded theory (Glaser and Strauss, 1967) 'contains some near mystical passages' (Melia, 1996).

What did the approach produce? The short answer is that it gave original and rich findings that were closely related to the data. It is this last point which gives the researcher such confidence – for each aspect of the substantive[4] theory produced, one can point to dozens of instances in the data which relate to it. Using the grounded theory gives confidence for other reasons too – a chain of analysis can be demonstrated in the research. Also, relating the substantive theory, once produced, to other theories increases the relevance and fit of that substantive theory.

From a personal perspective, I found the process of coding data and the emergence of concepts a creative and joyous process that also taught me a great deal about rigorous research. Grounded theory is by definition a rigorous approach – it demands time, it demands a chain of analysis, and the relating of findings to other theories. Finally, it demands trust—Glaser has called grounded theory method a 'delayed action' phenomenon, in that important findings sometimes emerge after one has lived with the data for a while (Glaser, 1992). The experience of doing grounded theory gave me a sensitivity to other theories, as the process of coding and considering relationships between codes makes it easier to think theoretically about other theories (Glaser, 1978).

PHILOSOPHICAL ISSUES

Turning now to philosophical issues associated with grounded theory, there are at least three issues that should be considered when a researcher decides to use the grounded theory approach. All of these issues had to be resolved in the research example given in the previous section, and an alert reader may already have some inkling of how they were resolved for that particular example. It seems appropriate here to examine these three issues in more detail.

1. What is the philosophical position of grounded theory itself? Does this matter if the researcher decides to use grounded theory as a technique within a given research paradigm?

2. Grounded theory has a long history, and has been extensively written about by its originators and followers (Glaser and Strauss, 1967; Glaser, 1978; Strauss, 1987; Strauss and Corbin, 1989, 1990; Glaser, 1992; Strauss and Corbin, 1994; Glaser, 1995, 1998). During this time it has evolved, differing advice about application can be found in different books, and most importantly,

there has been a major disagreement between the co-originators. Given this, can a researcher really claim to be using grounded theory if they use the 1990 Strauss and Corbin work as their sole guide, in the light of both a large body of work and this disagreement?

3. How does a researcher deal with the issue of theoretical sensitivity and what it implies? Does relevant literature have to be ignored until data is collected?

The Philosophical Position of Grounded Theory and Its Relevance for the Information Systems Researcher

Grounded theory method is claimed to have emerged from symbolic interactionism (Annells, 1996), something rarely remarked upon in information systems literature. Symbolic interactionism holds that the individual enters their own experience only as an object, not a subject, and that entry is predicated on the basis of social relations and interactions (Mead, 1962, in Annells, 1996). This leads Annells (1996) to place grounded theory within an ontology of critical realism, as part of the post-positivist paradigm (Guba and Lincoln, 1994). Critical realism holds that there is one reality, however imperfectly apprehendible (Guba and Lincoln, 1994). Annells (1996) points to statements by Glaser (1992) about the classic mode of grounded theory focusing on 'concepts of reality'(p.14) and searching for 'true meaning' (p.55) as evidence of a critical realist position. It should be noted however, that Glaser (1999) has said:

> In some quarters of research, grounded theory is considered qualitative, symbolic interaction research. It is a kind of takeover..

Both symbolic interactionism and grounded theory have also been claimed as *interpretive* approaches, where the ontology is one of socially constructed meaning systems which is based on an internal experience of reality (Neuman, 1997). Grounded theory has also been linked to philosophical hermeneutics which offer an alternative to empiricist and historicist accounts of science (Thompson, 1990). Strauss and Corbin's (1990) suggestion of a conditional matrix, which incorporates consideration of larger contextual issues of historical, political and economic conditions, can be seen to lean toward a relativist approach (Annells, 1996). Strauss (1987) suggests that the researcher is actively involved with the method, and again this can be interpreted as a relativist statement (Annells, 1996). In contrast to the very post positivist sounding 'criteria and canons' of Strauss and Corbin (1990) for judging grounded theory studies, Glaser (1992) states that the criteria for judging a grounded theory are 'fit, work, relevance, modifiability, parsimony and explanatory scope'. These sound considerably closer to Guba and Lincoln's (1989) *authenticity* criteria for constructivism, such as fairness and improved understanding, than Strauss and Corbin's (1990) criteria. So, it seems that Glaser, Strauss, and Strauss and Corbin have at different times, and sometimes simultaneously, leaned toward both post positivism and interpretivism. Madill, Jordan and Shirley (2000) argue quite convincingly that the philosophical position adopted when using grounded theory depends on the extent to which the findings are considered to be discovered within the data, or as the result of construction of inter-subjective meanings. They locate the former view as Glaser's (1992) position and the latter as Strauss and Corbin's (1990).

One of the enduring paradoxes of grounded theory is that, above all it is an *inductive* method and has been stated as such from the very beginning (Glaser and Strauss, 1967), and yet it is seen as a post positivist method. Post positivism like its predecessor still places great value on deduction as a way of discovering a research problem. Grounded theory's original aim was to inductively generate formal theory, via the route of substantive theory, in the field of sociology. Substantive theory pertains to a particular area, and the idea is that, at a certain level, it shades into bigger or more formal theories (Strauss 1987). Other theories pertaining to the same area as the substantive area need to be grappled with as competing analyses (Strauss, 1987).

Glaser (1992) lays great stress on the 'emergent' nature of grounded theory method, and states that the data should not be 'forced' into conceptual categories. It is the inductive and emergent nature of the method that seems most at odds with an underlying ontology of critical realism. Strauss and Corbin (1990) talk of a 'reality that cannot actually be known, but is always interpreted.'

Perhaps we should view the ontology of grounded theory method, as proposed in 1967, as being a product of the political and historical context of the time. The various indicators of philosophical position from the literature since may be seen as a product of more recent shifting ideas and epistemologies in qualitative research. Above all it is a *method*, and as such, can be used comfortably in most paradigms. For instance, in information systems it has been largely used within an interpretive context (Toraskar, 1991; Orlikowski, 1993; Urquhart, 1997, 1998a, 1999a, 1999b; Baskerville and Pries-Heje, 1999; Trauth, 2000) but also more positivist ones (e.g., Adams and Sasse, 1999)

It is probably appropriate to leave the penultimate word on this to Glaser (1999), who stated during a conference address:

"Let me be clear. Grounded theory is a general method. It can be used on any data or combination of data."

So, information systems researchers, in my opinion, can safely use the grounded theory approach to analyse qualitative data, irrespective of which research paradigm they are using. For instance, in the research example given earlier, I located my use of grounded theory within a research paradigm of constructivism. In common with most of my information systems colleagues, it was used primarily as a method of analysing data, sometimes set against other analyses (Toraskar, 1991; Orlikowski, 1993; Urquhart, 1997, 1998a, 1999a, 1999b; Baskerville and Pries-Heje, 1999; Adams and Sasse, 1999; Trauth, 2000).

What Was the Disagreement Between Glaser and Strauss About, and What Difference Does it Make to the Information Systems Researcher?

Glaser described the book written by Anselm Strauss and Juliet Corbin in 1990 thus:

"It distorts and misconceives grounded theory, while engaging in a gross neglect of 90% of its important ideas" (Glaser 1992, p 2).

So, clearly this is a major disagreement between the cooriginators, rather than a minor point of technical application of grounded theory method. He felt so strongly about the book that he requested it be pulled from publication, and when it was not, wrote a correctional rejoinder to the Strauss and Corbin book, called "Emergence vs. Forcing: Basics of Grounded Theory Analysis" (Glaser, 1992). This book followed the headings of Strauss and Corbin, and wrote a rejoinder or correction for each section. Glaser's basic problem with the Strauss and Corbin book is hinted at in the title of his rejoinder. He felt that it was about what he called 'forced, conceptual description" rather than allowing the analysis of the data to emerge naturally ("emergence") through constant comparison, which is the process of comparing whatever data bit is being analysed against both the chief concern of the participants and the emergent categories.

Strauss and Corbin (1990) recommend the use of a 'coding

paradigm' which makes connections between data categories. This coding paradigm (modified from Strauss's 1987 coding paradigm) suggests that the researcher looks for context, conditions, action/ interactional strategies, intervening conditions and consequences as a guide to establishing these relationships. Glaser (1992) had this to say about the use of the coding paradigm:

> "If you torture the data long enough, it will give up! This is the underlying approach in forcing preconceptions of full conceptual description. The data is not allowed to speak for itself as in grounded theory, and to be heard from, infrequently it has to scream. Forcing by preconception constantly derails it from relevance." (Glaser 1992, p 123)

Glaser pointed out that this coding paradigm was a variation of only one of a possible 18 theoretical coding families he had put forward previously (Glaser, 1992, 1978). Therefore to 'force' all axial coding through one paradigm is not grounded theory, rather conceptual description, and ignores the emergent nature of grounded theory. For instance, Glaser (1992) makes the point that, if a category is a condition of a property, then it will emerge as such. Certainly this has been my own experience—categories have to be reorientated and some reconsidered as either properties of other categories, or indeed relationships (Urquhart 1998b). Similarly, I found Strauss's (1987) coding paradigm rather restrictive and instead used it as a jumping off point to think about categories, as opposed to a coding guide in of itself, as illustrated earlier in this chapter. Other grounded theorists too, have found the Strauss and Corbin (1990) book rather formulaic and overburdened with many rules (Melia, 1996; Kendall, 1999).

It is indeed possible that the two cooriginators had different ideas about grounded theory right from the inception of grounded theory method – Melia (1996) quotes Stern (1994), asserting that "students of Glaser and Strauss in the 1960s and 1970s knew that the two had a quite different modus operandi, but Glaser only found out when Strauss and Corbin's *Basics of Qualitative Research* came out in 1990". This would seem to be borne out by Glaser's (1992) remark:

> "What has started out as a book of corrections ended up showing that Strauss indeed has used a different methodology all along, probably from the start in 1967" (Glaser 1992, p.122).

So what relevance does this disagreement have for information systems researchers?

Firstly, it is important from a scholarly perspective to *fully* understand the body of grounded theory work, rather than import *one* view

(Strauss and Corbin 1990) which has been said to be unrepresentative by the cooriginator of grounded theory (Glaser, 1992). It follows then, that researchers need to be aware of what version they use, and Kendall (1999) is very helpful in illustrating the differences and the impact of adopting one version or the other. My own experiences have led me to prefer the Glaserian version, but each researcher needs to decide this for her or himself, and fully inform themselves on the debate if intending to use grounded theory.

Secondly, it may be that information systems researchers find grounded theory method difficult to use precisely because the later books (Strauss, 1987; Strauss and Corbin, 1990), in their attempt to make grounded theory more accessible, end up putting forward many procedures, strictures and injunctions. As such, these books obscure what is a fairly simple and useful idea for coding data, that of *constant comparison*, or conceptualising patterns among incidents (Glaser, 1992):

> "Using constant comparison method gets the analyst to the desired 'conceptual power', with ease and joy. Categories emerge upon comparison and properties emerge upon more comparison. And that is all there is to it" (Glaser 1992, p.43).

Finally, the disagreement helps us understand something very fundamental about coding data—the difference between allowing the data to speak to us and coding from the ground up, rather than imposing preconceived categories, which is a top-down method of coding.

What is 'Theoretical Sensitivity' and What Are the Implications for the Researcher's Relationship With the Literature?

One of the oft quoted misconceptions about grounded theory is that the researcher does not do any literature searching and goes into the field 'blind,' as it were. This means that, in some departments and schools of information systems, a grounded theory study seems not to be an option as it does not fit into the standard deductive template of 'do your literature search first, then go out into the field'. The idea that a researcher does not refer to literature is a corruption of the idea of *theoretical sensitivity* (Glaser, 1978) and is entirely unhelpful as it conveys a false impression of grounded theory as unrigorous (it is quite the reverse) and naïve (which is not). Nevertheless, if using grounded theory method, the researcher has to relate to literature in a slightly different way to a conventional researcher, and this is worth discussing at some length, given how much confusion seems to surround this particular issue.

The concept of theoretical sensitivity was developed in 1978 by Glaser and was put forward to help researchers relate their analysed categories into theory (Glaser, 1978). One of the tenets of theoretical sensitivity is that the researcher enters the research setting with as few predetermined views as possible, especially logically deducted, prior hypotheses (Glaser, 1978). That said, theoretical sensitivity is increased by being steeped in the literature and associated general ideas (Glaser, 1978), so that a researcher will understand what theory is. Thus the idea of theoretical sensitivity can be seen as an injunction against a deductive mode of thinking rather than an injunction against the literature per se.

Glaser (1992) further elaborates by stating that theoretical sensitivity is:

"An ability to generate concepts from data and relate them to the normal models of theory in general" (Glaser 1992 p.31).

So, literature is used to help build the theory, and the substantive theory related to the literature, *but only once the substantive theory has been developed.* According to Glaser (1992), the dictum in grounded theory is that there is no need to review the literature in the substantive area under study, and that this idea is:

"...brought about by the concern that literature might contaminate, stifle or contaminate or otherwise impede the researcher's effort to generate categories..." (Glaser 1992 p.31).

He hastens to add though, that this applies only in the beginning, and that when the theory is sufficiently developed, that the researcher needs to review the literature in the substantive field and relate that literature to their own work (Glaser, 1992).

It is also worth relaying Strauss's (1987) opinion on this issue, who says that the advice about delaying the scrutiny of related literature applies full force to inexperienced researchers, but less so to experienced researchers who are already good at subjecting a theoretical statement to comparative analysis, and would question whether it would hold 'true' under different conditions.

CONCLUSION

This chapter has examined both practical and philosophical aspects of applying grounded theory method in the field of information systems. It seems appropriate here to conclude with some summary points about the application of grounded theory that have been covered in this chapter.

First, the somewhat ambiguous philosophical position of grounded theory method should give researchers pause to think. As it is an inductive, emergent method that is located mainly in post positivism, this means that researchers need to carefully consider their own philosophical position. That said, it is primarily a method, and can be used in several different paradigms. Researchers can choose not to take on any of the philosophical baggage that the body of grounded theory literature has, but they can only do this if they understand that literature.

Secondly, from a practical point of view, when using grounded theory method, I would suggest the following rules of thumb:

i.   Emergent theory comes from relationships between codes, so paying attention to this aspect pays dividends.

ii.   It is easy to fall into the trap, when coding, of simply labelling what you see rather than gaining any analytic insight. Dey (1999) is extremely helpful on this issue of labelling. The process of constant comparison helps alleviate this, as does consideration of theoretical codes such as Glaser's (1978). This is probably a better option than following the coding paradigms recommended by Strauss (1987), and Strauss and Corbin (1990), purely from the point of view of providing more flexibility.

iii. You should be able to clearly demonstrate a chain of analysis (however iterative the actual process was) from the open codes up.

iv.   Be flexible about naming of codes and categories, as the naming deeply affects how you and others see that concept.

v.   Be aware that properties can be elevated to categories, and that this might also apply to a relationship between categories.

vi.   Be prepared to live with your data a long time, and to do comparisons across data sources.

vii. Write theoretical memos and draw diagrams to spur thinking and creativity.

One final issue worth considering, albeit briefly, is what might constitute a 'good' study in information systems using grounded theory method. How could we evaluate a grounded theory study in information systems? Strauss and Corbin (1989) published a set of canons and criteria as to what a good grounded theory study might look like. There are a few difficulties with this set of canons and criteria, not least Glaser's (1992) critiques of both. He labelled the canons as 'canons for judging quantitative method research inappropriately applied to grounded theory', and commented that the criteria simply went through the grounded theory steps to see if they were done properly, rather than assessing fit and relevance of the concepts produced by those steps. Given the points made earlier about the ambiguous philosophical position occupied by grounded theory method, it might be difficult for researchers using an interpretive philosophy to fully comply with the canons and criteria.

Interpretive information systems researchers now have a set of their own criteria to deal with—that of Klein and Myers (1999). It has been my experience that the Klein and Myers principles, do, in the main, work with a grounded theory study using an interpretive paradigm, though the principle of multiple interpretations is harder to satisfy if one's aim is generating a substantive theory. The key to this not-so-obvious fit is the fundamental principle of the hermeneutic circle proposed by Klein and Myers (1999). It has already been mentioned in this chapter that parallels have been drawn between grounded theory method and philosophical hermeneutics (Thompson 1990). At a practical level, this can be seen in the constant comparison method adopted by grounded theorists when coding data.

One objective of this chapter has been to provide some guidance for information systems researchers using grounded theory for the first time, especially if they do not have access to an individual who has used grounded theory, which is more likely than not. Melia (1996) mentions the idea of 'minus mentoring', where the grounded theorist has learned the method from a book. Certainly in some circles this is seen to amount to an erosion of the ideas of grounded theory. I would not agree that it is impossible to learn grounded theory from a book, but it is extremely helpful to the beginner to participate in coding seminars where they can learn to interpret and conceptualise the data. Of the body of grounded theory, Strauss's 1987 book is extremely helpful to the beginner as it provides a coding seminar and also answers frequently asked questions.

As a (still) new discipline, we need to increase our mentoring in all kinds of qualitative research endeavours, and I would cite the production of books such as this as giving valuable guidelines to information systems researchers. From the point of view of learning how to 'do' grounded theory, there is still no substitute for sitting in a room with colleagues and exploring how data might be coded.[5]

My aims in writing this chapter were twofold: first, to offer some practical guidelines to information systems researchers interested in using grounded theory. Secondly, to clear up some common misconceptions about grounded theory method in information systems research and to consider some philosophical issues of application. My own experience as a 'minus mentee' has been by turns exasperating, exciting and deeply fulfilling. I am hopeful that use of grounded theory method in information systems will grow, as it provides a useful route to provide substantive and formal theory in information systems, and produces rich, interesting and above all, *relevant* findings which have an inescapable tie to the data.

APPENDIX 1 - ANALYTIC MEMO

Agenda Setting as a key to both Conceptualisation and Tactics   *AM 5297*

The purpose of this memo is to try and clarify a few thoughts on *agenda setting*. Agenda setting has many elements, both conceptual and tactical. It could be defined as the process by which a participant (generally the analyst) sets out the topic for discussion, and sometimes the process for managing that topic. Another way of viewing agenda setting is that it comprises a framework for conceptualisation and negotiation (which is a tactical element). Who actually sets the agenda for discussion gives some indicators as to the type of relationship between the analyst and client (cf Hirschheim's four models). There is evidence in negotiation literature that whoever sets up the framework for discussion is at a tactical advantage.

The way the topic is introduced gives many clues as to how the participant is conceptualising the problem. Therefore by looking at how the analyst defines the problem, we can gain insight into the conceptual schema the analyst is using. What is also of interest is if this conceptual schema influences the solution proffered in the conversation. More broadly, the notion of a conceptual schema that the analyst employs can be seen to be important in the design of information systems. For instance, if the problem is narrowly defined by virtue of the conceptual schema, then the resultant design may be similarly narrow in scope. As the design of information systems rests purely on concepts, then the conceptual schema used becomes very important.

In addition, by examining how the client presents the problem, one can judge if differing schemas are bridged in a joint conceptualisation. If analysts recognise the schemas they are applying to an information system, then they can perhaps apply one or a number of schemas that are appropriate for the problem. It may be that bringing in a too rigid conceptual schema limits the solution, and that broader schemas are appropriate. It may also be that a tactic of information gathering, without bringing in a particular schema, might be more successful.

Agenda setting can be seen as a mediating process between tactics and concepts. As such it could be construed as a relationship. It also provides a bridge between structure of the text and the social processes evidenced by the text, thus helping to resolve the structural/processual dichotomy encountered when analysing discourse. As agenda setting contains both conceptual and tactical elements, one can deduce from the text: the concepts that are informing tactics; how the problem is formulated influences tactics; how the tactics used by both participants influence joint conceptualisation.

Possibly agenda setting is the core category of the study – that process of *how* analysts and clients reach agreement (which after all is the research question). Although the term agenda setting implies a starting point, communications research has put forward the notion of topic as a chain of subtopics – this also fits in neatly with the idea of evolving conceptualisation. The rest of this memo will give instances of agenda setting and its elements, and will discuss how it might play a role in linking concepts and tactics.

APPENDIX 2 - INTEGRATIVE DIAGRAM

(ID 29/9/96)

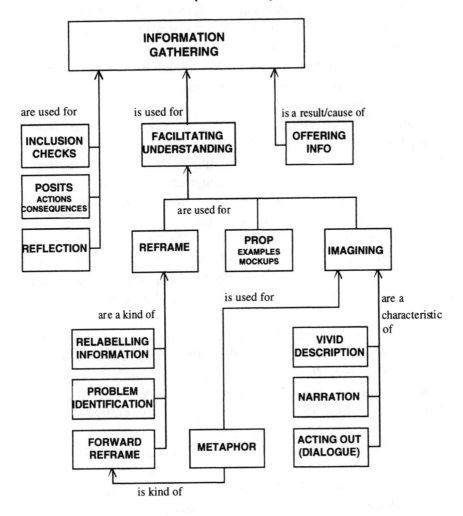

REFERENCES

Adams A and Sasse MA (1999). Users Are Not The Enemy, *Communications of the ACM*, December.

Annells M (1996). Grounded Theory Method: Philosophical Perspectives, Paradigm of Inquiry, and Postmodernism, *Qualitative Health Research*, 6(3), Sage: CA.

Baskerville R and Pries-Heje J (1999). Grounded action research: a method for understanding IT in practice, *Accounting, Management and Information Technologies*, 9(1).

Blumer H (1969). *Symbolic Interactionism: Perspective and Method*, Prentice Hall: Englewood Cliffs, NJ.

Candlin CN (1985). Preface to Coulthard M , *An Introduction to Discourse Analysis*, 2nd Edition, Longman, Essex UK.

Cresswell J.W (1998). *Qualitative Inquiry and Research Design: Choosing Among Five Traditions*, Sage:London

Dey I (1993). *Qualitative Data Analysis*, Routledge:London.

Dey I (1999). *Grounding Grounded Theory*, Academic Press: CA.

Glaser, B.G and Strauss A. (1967). *The Discovery of Grounded Theory: Strategies for Qualitative Research*, Aldine Publishing Co:Chicago, IL.

Glaser B.G (1978). *Theoretical Sensitivity: Advances in the Methodology of Grounded Theory*, Sociology Press: Mill Valley, CA.

Glaser B.G (1992). *Basics of Grounded Theory Analysis: Emergence vs. Forcing*, Sociology Press: Mill Valley, CA.

Glaser B.G (1995). (Ed), *Grounded Theory, 1984-1994*, Vols 1-2, Sociology Press: Mill Valley, CA.

Glaser B.G (1998). *Doing Grounded Theory: Issues and Discussions*, Sociology Press: Mill Valley, CA.

Glaser B.G (1999). The Future of Grounded Theory, *Qualitative Health Research*, 9(6), Sage:CA, 836-845.

Guba E.G and Lincoln Y (1989). *Fourth Generation Evaluation*, Sage: CA.

Guba E.G and Lincoln Y (1994). Competing paradigms in qualitative research in Denzin N.K and Lincoln Y.S eds (1994) *Handbook of Qualitative Research*, Sage:CA, 105-116.

Guinan PJ (1988). *Patterns of Excellence for IS Professionals – An Analysis of Communication Behaviour*, ICIT Press, Washington.

Kendall J (1999). Axial Coding and the Grounded Theory Controversy,

Western Journal of Nursing Research, 21(6), Sage:CA, 743-757.

Klein H.K, Nissen H-E and Hirschheim R (1991). A Pluralist Perspective of the Information Systems Research Arena, *Information Systems Research: Contemporary Approaches and Emergent Traditions*. H.-E. Nissen, H.K. Klein and R. Hirschheim (eds.) Amsterdam: North-Holland.

Klein H.K and Myers M.D (1999). A Set of Principles for Conducting and Evaluating Interpretive Field Studies in Information Systems, *MIS Quarterly*, Special Issue on Intensive Research, 23(1).

Madill A, Jordan A, and Shirley C (2000). Objectivity and Reliability in Qualitative Analysis: Realist, Contextualist and Radical Constructionist Epistemologies, *British Journal of Psychology*, Vol 91, 1-20.

Markus M.L (1997). The Qualitative Difference in Information Systems Research and Practice, Lee AS, Liebenau J and DeGross JI (Eds) *Information Systems and Qualitative Research*, Chapman and Hall, London.

Mead G.H (1962). *Mind, Self and Society*, (C.Morris, Ed), University of Chicago Press: Chicago.

Melia K.M (1996). Rediscovering Glaser, *Qualitative Health Research*, 6(3), Sage:CA.

Miles MB and Huberman AM (1994). *Qualitative Data Analysis – An Expanded Sourcebook*, Second Edition, Sage:CA.

Myers M (1997). Interpretive Research in Information Systems, in J Mingers and F Stowell (Eds), *Information Systems: An Emerging Discipline?*, McGraw-Hill: London, 239-266.

Myers M.D and Walsham G. (1998). Exemplifying interpretive research in information systems: an overview, *Journal of Information Technology* 13(4), 233-234.

Neuman W L (1997). *Social Research Methods*, 3rd Edition, Allyn and Bacon.

Orlikowski WJ and Baroudi JJ (1991). "Studying Information Technology in Organizations: Research Approaches and Assumptions, *Information Systems Research*, 2(1), 1–28.

Orlikowski WJ (1993). CASE Tools as Organizational Change: Investigating Incremental and Radical Changes in Systems Development, *MIS Quarterly*, 17(3).

Pettigrew AM (1985). "Contextualist Research and the Study of

Organizational Change" in Mumford E, Hirschiem R, Fitzgerald G, and Wood-Harper T (Eds), *Research Methods in Information Systems*, Elsevier, North Holland.

Schultze U (2000). A Confessional Account of an Ethnography About Knowledge Work, *MIS Quarterly*, Special Issue on Intensive Research, 24(1).

Schön D.A (1983). *The Reflective Practicioner: How Professionals Think in Action*, Basic Books, NY.

Spradley J.P. (1979). *The Ethnographic Interview*, Harcourt Brace Jovanovich College Publishers, Fort Worth.

Stern P.N. (1994). Eroding Grounded Theory, in J.Morse (Ed), *Critical Issues in Qualitative Research Methods*, Sage:CA, 212-223.

Strauss A (1987). *Qualitative Analysis for Social Scientists*, Cambridge University Press: Cambridge.

Strauss A and Corbin J (1989). Grounded Theory's Applicability to Nursing Diagnostic Research, In Monograph of the *Invitational Conference of Research Methods for Validating Nursing Diagnoses*, Palm Springs, CA, 4-24.

Strauss A and Corbin J (1990). *Basics of Qualitative Research*, Sage:CA

Strauss A and Corbin J (1994). Grounded Theory Methodology – An Overview in Denzin NK and Lincoln YS (Eds) *Handbook of Qualitative Research*, Sage: CA.

Tan, M (1989). "An Investigation into the Communication Behaviours of Systems Analysts", *Unpublished PhD thesis*, University of Queensland.

Thompson, J (1990). Hermeneutic Enquiry, In E.Moody (Ed), *Advancing Nursing Science Through Research*, Sage:London, pp223-280

Trauth, E.M. (2000). *The Culture of an Information Economy: Influences and Impacts in the Republic of Ireland.* Dordrecht, The Netherlands: Kluwer Academic Publishers.

Toraskar K.V (1991). How Managerial Users Evaluate Their Decision-Support: A Grounded Theory Approach *Information Systems Research: Contemporary Approaches and Emergent Traditions*. H.-E. Nissen, H.K. Klein and R. Hirschheim (eds.) Amsterdam: North-Holland.

Trauth, E.M. and Jessup, L. (2000). "Understanding Computer-mediated Discussions: Positivist and Interpretive Analyses of Group Support System Use." *MIS Quarterly*, Special Issue on Intensive

Research, 24(1).

Urquhart, C. (1997). Exploring Analyst-Client Communication: Using Grounded Theory Techniques to Investigate Interaction in Informal Requirements Gathering in Lee, A. S., Liebenau, J., and DeGross, J. I. (Eds.) *Information Systems and Qualitative Research*. London: Chapman and Hall.

Urquhart, C .(1998a). Analysts And Clients In Conversation: Cases In Early Requirements Gathering, *Proceedings of the Nineteenth International Conference on Information Systems,* (Eds) Hirschheim R, Newman M, DeGross J.I.

Urquhart C (1998b). Strategies for Conversation and Systems Analysis in Requirements Gathering: A Qualitative View of Analyst-Client Communication, *The Qualitative Report*, 4(1/2), March/July 1998, http://www.nova.edu./ssss/QR/QR4-1/urquhart.html, ISSN 1052-0147.

Urquhart C (1999a). Reflections On Early Requirements Gathering – Themes From Analyst-Client Conversation, in Larsen T.J, Levine L, DeGross J.I (Eds) *Information Systems: Current Issues and Future Changes,* IFIP: Laxenburg, Austria.

Urquhart C (1999b). Themes In Early Requirements Gathering – The Case Of The Analyst, The Client And The Student Assistance Scheme, *Information Technology and People,* 12(1).

Walsham G (1995). The Emergence of Interpretivism in in Information Systems Research, *Information Systems Research*, Volume 6, Number 4, 376–394.

Walsham G and Sahay S (1999). GIS for District-Level Administration in India: Problems and Opportunities, *MIS Quarterly*, Special Issue on Intensive Research, 23(1).

ENDNOTES

1 Using Orlikowski's and Baroudi's (1991) definition, ontology consists of beliefs about the nature of physical and social reality.

2 Again using Orlikowski and Baroudi's (1991) definition, epistemology consists of beliefs about knowledge.

3 While I could have equally called it a 'summary', the word 'vignette' reinforced the narrative approach.

4 Substantive in the sense of pertaining to the area of investigation (Strauss 1987).

5 Interested readers might like to know that Glaser has set up a
mentoring system, which takes place over the Internet and is called
'minus mentoring plus one' (www.groundedtheory.com).

Chapter VI

Doing Critical IS Research: The Question of Methodology

Dubravka Cecez-Kecmanovic
University of Western Sydney, Australia

INTRODUCTION

Critical information systems (IS) research denotes a critical process of inquiry that seeks to achieve emancipatory social change by going beyond the apparent to reveal hidden agendas, concealed inequalities and tacit manipulation involved in a complex relationship between IS and their social, political and organisational contexts. It has its philosophical and theoretical roots in *critical social theory* (Held, 1980; Fay, 1987; Morrow and Brown, 1994). As a critical social researcher studies the social life of people in order to help them change conditions and improve their lives, so too does a critical IS researcher. By demystifying technological imperatives and managerial rationalism justifying a particular information system design, the critical IS researcher helps both IS practitioners and users understand its social consequences, envisage desirable alternatives and take action.

Like interpretive approaches to IS research, critical theory-informed approaches came along as a reaction to positivism. While interpretive researchers aim to understand and describe "the *context* of the information system, and the *process* whereby the information system influences and is influenced by its context" (Walsham, 1993), critical IS researchers go further to expose inherent conflicts and contradictions, hidden structures and mechanisms accountable for these influences. Critical IS researchers aim to reveal interests and agendas of privileged groups and the way they are supported or protected by a particular information system design or use. More generally, they aim to discover and expose attempts to design and (mis)use IS to deceive, manipulate, exploit, dominate and disempower people. By doing so they aspire to help them resist these attempts, hinder such misuse of IS and promote liberating and empowering IS design and use.

Such concerns and critical orientations have inspired diverse research programs in IS that have been recognised as a new, critical paradigm in IS research (Hirschheim and Klein, 1989; Hirschheim et al., 1996; Iivari et al., 1998). While they succeeded to (re)open the fundamental questions of the nature of IS and their social reality, and increase awareness of normative knowledge, critical IS researchers faced serious problems. The very assumptions of critical IS research have been questioned and its objectives deemed unachievable. Critical researchers have even been accused of promoting yet another 'totalising discourse' (Wilson, 1997). On the other hand, proponents of the critical paradigm in IS have long been aware of its weak empirical grounding and the lack of appropriate empirical methods (Lyytinen and Klein, 1985; Lyytinen, 1992; Klein, 1999). Most notably, critical researchers themselves identified a problematic relationship between critical theory and empirical research methods as a key problem, though not unique to the IS research (Morrow and Brown, 1994, Klein, 1999; Forester, 1992).

This chapter addresses these issues by focusing on the question of methodology in critical theory-informed IS research. Methodology is understood here in its philosophical sense as an overall strategy of conceptualising and conducting an inquiry, and constructing scientific knowledge. Methodology, therefore, refers not only to research methods or techniques (such as case study or interview), but also to the epistemological assumptions of methods and how they are linked to a particular theory. A critical research program in IS sets an agenda and the types of explanatory substantive problems for which some methods are more appropriate than others. Critical research methodology is explicitly concerned with the choices about linking theories and methods in any specific research context. Moreover, the ultimate concern of critical IS research methodology is the implication of critical inquiry on social practices in the development and use of IS.

The purpose of the chapter is to explore the relationships between a critical theory, empirical methods and research questions asked in IS research situations. Given the diversity of contemporary critical research programs such an exploration may involve a comparative study across different programs and theories or may focus on specific ones with limited generalisations. In this chapter I endeavour the latter. I examine a methodological strategy I applied to link a particular theory, specifically Habermas' theory of communicative action (1984, 1987), with empirical methods in an inquiry of organisational public discourse via Computer-Mediated Communication. By drawing on my experience from this inquiry I reflect on the critical issues I faced and methodological choices I made. While such an account is necessarily idiosyncratic, I attempt to demonstrate that the lessons learned are relevant for the emerging debate on methodological issues in critical IS research.

AN EXAMPLE OF CRITICAL IS RESEARCH: A STORY OF THE UNIVERSITY DISCOURSE VIA COMPUTER-MEDIATED COMMUNICATION

This section aims to provide an illustrative example of critical field research into organisational discourse via computer-mediated communications (CMC) during a university restructuring process. The intention, given the methodological focus of this chapter, is not to substantively elaborate on the field study itself, which has been presented elsewhere (Cecez-Kecmanovic et al. 1999, 2000; Treleaven et al. 1999), but to describe it briefly to illustrate the type of questions asked and to reflect on the methodological issues and strategies.

As a member of University X in 1996-1998 I conducted a field study of a University-wide strategic restructuring process triggered by a crisis following funding cuts by the Federal Government. At the end of 1996, the President of University X initiated a *consultation with staff* with the aim to involve staff[1] in a broad-based discussion about the challenges the University was facing and future potential restructuring and rationalisation of the University to respond to these challenges. The consultation took place face-to-face (in public forums, facilitated

workgroups, a conference, and a variety of working teams and committees) and via CMC. CMC consisted of a particular set-up of e-mail and
intranet managed by a coordinator and accessible by all staff. It enabled
each staff member to get, send and retrieve messages and documents
related to the consultation. As the consultation process evolved throughout 1997, the discussion via CMC emerged as a key University discourse
that ultimately impacted upon the decision how to restructure the
University.

It is important to note that the purpose of CMC, as expressed by the
President and the coordinator on various occasions, was to:

- enable organisation-wide communication, discussion and sharing
  of information independent of limitations imposed by time and
  space.
- maintain an accessible electronic repository of all messages and
  documents created in the process.
- enable effective and efficient coordination between different
  individuals and groups involved in the consultative process.

Throughout the consultative process, more than 130 messages,
discussion papers and documents (to be referred to as e-mail messages)
were exchanged via CMC. Approximately 30% of those were submitted during initial exploratory phases before September 1997. The
President then proposed (by e-mail) his draft *Restructure* document,
outlining major changes of the academic structure, funding, income
generation, administration and management. He clearly stated that he
wanted to hear staff responses and that all issues were open for
discussion. The ensuing so-called *September discussion* included 67 e-
mail messages (by individuals, schools and faculties) ranging from one
paragraph to 15 pages. Shortly after the close of this discussion the
President announced his final *Restructure and Implementation Plan*
document which was subsequently implemented. The rest of the
discussion till the end of 1997 amounted to less than 20% of the
messages.

The extensive use of CMC throughout the consultation made all
information readily accessible to all staff, opened the restructuring
process to all interested parties, and generally gave an impression that
the process was transparent and inclusive. All through the consultation
until September, the expectations that CMC would foster freedom of
speech, increase equality of participation and reduce status-related
barriers seemed to be well-founded. However, as the electronic discus-

sion unfolded, especially during the intensive September discussion, participants noticed that the President did not actually engage in argumentation. He did not respond to participants' criticism of and arguments against the proposed changes in his *Restructuring* document, nor did he comment on counter-arguments and alternative proposals. His understanding of the University problems (e.g., how funding cuts can be absorbed and what makes the academic structure flexible) and his major solutions (e,g., centralisation of staff funding) were not altered despite well-argued criticism and counter-proposals expressed by many participants during the debate. All major changes he proposed in the draft *Restructuring* document remained unchanged in the final document, and subsequently implemented.

My interest in this study initially was the examination of transformational power of CMC in the University-wide discourse, especially the potential for empowerment and democratisation. Early in 1997 I got together with two other members of the University, one academic and another member of the staff development unit, and we created a small research team. We broadened the focus of the field study to include social, cultural, historical and political context (at both local, group level and University level) and the ways CMC was appropriated in different contexts. As we were members of the University we got all the e-mails from the consultative process. We stored them and analysed them as the process evolved. We also observed meetings and other face-to-face discussions and kept field notes. We regularly exchanged our observations and often talked about problems of being both participants and observers in the process.

The interpretation of e-mail messages and documents, and the analysis of the flow of argumentation process, became the central focus of our research strategy. As the consultation evolved, growing more complex and diverse, we became increasingly concerned with differences in our interpretation of discussions and our understanding of others, and what was going on through CMC. That, among other things, prompted us to consider interviewing staff in order to get more information and test some of our interpretations. In the second half of 1997 we started semi-structured interviews with staff. A new team member (a former employee of the University) was engaged to help us conduct interviews (in most cases we interviewed individual staff but in a few cases a small group). By mid-1998, 50 interviews with a wide range of University members were completed, taped and transcribed.

The President's establishment of the consultative process, including CMC as an extended social space for public discourse, on the one hand, and subsequent ignorance of criticism and counter-arguments by staff, on the other, were contradictory. It was not until September, when the most intensive debate about *Restructuring* took place, that I became aware of this sharp contradiction. It was difficult to explain why the President would invest so much effort and energy to *consult with staff* if he ignored their criticism, arguments and proposals (I'll refer to this as the first contradiction). Moreover, how could such an open, transparent and seemingly democratic process, assisted by CMC, be used to advance a repressive outcome (the second contradiction)?

REFLECTIONS ON METHODOLOGY

In the above example methodological choices were made along the way, as research evolved and questions emerged. My experience confirms that "a defining characteristic of critical research methodology is that choices about linking theories and methods are an ongoing process that is contextually bound, not a technical decision that can be taken for granted through references to the 'logic of science' " (Morrow and Brown, 1994, p. 228). My reflections in this section will highlight some aspects of *reflexivity* and *dialectical character* of the critical IS research methodology, as revealed in the relationship between theory and method and between a theory and methodological strategies.

Reflexivity

Research in the critical tradition is characterised by reflexivity, involving forms of self-conscious criticism as part of a strategy to conduct critical empirical research. Researchers explore their own ontological and epistemological assumptions and preferences that inform their research and influence their engagement with a study. By intentionally expressing, questioning, and reflecting upon their subjective experiences, beliefs, and values, critical researchers expose their ideological and political agendas. Thus, as Kincheloe and McLaren imply, "critical researchers enter into investigation with their assumptions on the table, so no one is confused concerning the epistemological and political baggage they bring with them to the research site" (2000, p. 292).

In the context of critical research methodology such reflective practice is relevant in at least three aspects. Firstly, it helps researchers understand their own engagement with subjects in a field, identify sources of different/conflicting views and beliefs, and potentially change their own. Secondly, it helps researchers make explicit connections and comparisons with relevant circumstances and experiences in the past (in the same or different organisation). Thirdly, it enables a team of researchers to develop mutual understanding and explore differences in interpretation and explanation of empirical material.

The assumptions and beliefs I brought into the research of CMC date back to the 1970s, years of great hope for democratisation in my former country (Yugoslavia). By a stroke of luck I was introduced first-hand to the contemporary critical thought by attending the famous Summer school on the island of Korcula that attracted leading critical theorists from both West and East[2]. As a student (and an active participant in the student movement), I was very appreciative of these ideas and came to believe that they would lead towards greater democratisation of our socialist system.

Later on, as an academic and researcher in IS, I conducted a large 5-year project on "The Social System of Information" (in the then Republic of Bosnia and Herzegovina) that had a dual objective: to analyse and assess the existing information production and decision-making processes in society, and to propose a new design of this system to better serve the needs of the self-management system from the local community level to the Parliament (Cecez-Kecmanovic, 1987). This project was based on the assumptions that the accessibility of free (and uncontrolled) sources of information to citizens and their representatives in the Parliament and other social and political institutions will support and enable democratisation processes. The proposed design relied on the advanced information technology (IT) already available in some institutions (e.g., statistical organisations, unions). But the break-up of Yugoslavia and a tragic course of events in the last decade prevented its implementation. My beliefs and assumptions about the transformative potential of IS remained unchallenged.

Moreover, my beliefs in a democratising potential and emancipatory role of IS were being substantiated by the literature on the impacts of groupware and CMC (Kiesler and Sproull, 1992; Siegel et al., 1986). Claims such as that CMC technologies "are surprisingly consistent with Western images of democracy" (Sproull and Kiesler, 1991, p.13) seemed to me reasonably well grounded.

It is not surprising, then, that I did not have any doubts in my mind that the expected (and publicly declared) role and purpose of CMC in the University consultative process would be realised. In fact, my beliefs and assumptions played an important role in my framing the research questions and choosing strategies to investigate it and conduct research. It was in the course of this research that I started to question my assumptions and beliefs, and to ask myself about their origins. Needless to say that in the course of this research, I learned to articulate, question, and examine my assumptions and beliefs. I also learned to disclose my illusions, values or ideologies. "The point is—as Garrison (1996) explains—not to free ourselves of all prejudice, but to examine our historically inherited and unreflectively held prejudices and alter those that disable our efforts to understand others and ourselves" (p. 434).

Another moment of self-reflexivity became apparent when we (the members of the research team) attempted to 'resolve' our differences in interpreting meanings (what participants actually *meant* by saying), in understanding their actions (what they actually *did* by saying), and in formulating our own actions (what can or should *we do* to assist participants). While I was conscious that our interpretations were grounded in our lifeworld and our individual life histories, experiences and prejudices, I had difficulties in finding strategies to overcome it. Denying or suppressing our individual differences, however, would have distorted our communication and collaborative process. Self-reflexivity, we found, is critically important but does not magically resolve differences. Continued questioning of individual assumptions and interpretations, as well as criticism and self-criticism, seemed at times disruptive to our otherwise good team relationships and cooperative spirit. Living with differences, nurturing critical self-reflection and inter-subjective reflectivity as we went, turned out to be our way of 'resolving' them. As a result we did not necessarily eliminate our differences ,but we certainly enriched and enlarged the set of our shared assumptions, beliefs and values, that is, our assumed *background knowledge*.

While these reflective practices may be more or less important for any empirical research strategy, they are of fundamental importance to the critical research strategy. As truth is not believed to be 'out there' waiting to be discovered, no one has a monopoly on correct interpretation. Critical research methodology is explicitly concerned with the issues of reflexivity, linking them to the more visible issues of empirical methods and strategies of inquiry, that I discuss in the following sections.

Relationship Between Theory and Method

That the choice of methods and the ways they are used to conduct research cannot be separated from the theory informing the research and the problems to be investigated, has been often emphasised (Morrow and Brown, 1994). Some empirical methods are more or less appropriate in terms of their use to answer particular questions posed by a critical IS research program. Moreover, a particular research context and conditions for conducting research limit both the choice of methods and the way they are used in practice. While it may look quite simple and straightforward, the relationship between theory and method is rarely so in any concrete research circumstance. By referring to the study of CMC in the University consultation process, I will point to a moment when this relationship became problematic and needed to be reconsidered as part of the changed research strategy.

The purpose of the study initially was to explore, document, and interpret the use and appropriation of CMC in the consultative process focusing on different social and cultural contexts (within groups and at the University level). Initially the study was conceptualised as an interpretive ethnography (Schwandt, 2000). We engaged in participant observation of face-to-face meetings and other less formal gatherings. In addition we downloaded all messages and documents submitted by CMC as part of the consultative process. We analysed the content of these texts and interpreted the meanings to gain understanding of the consultation and how CMC was appropriated. Given our initial research question and the hypothesis that CMC will be appropriated differently depending on the social context and culture, ethnographic methods seemed appropriate. The observation of local contexts (both academic and administrative departments) and the ways CMC was used to communicate the views of members in these departments provided rich descriptions we subsequently analysed and reported (Cecez-Kecmanovic et al., 1999).

As the consultation evolved, the discussion via CMC intensified and especially during September took over and became the major public discourse concerning the future of the University. As participants in this process, we sensed some major changes in the significance of CMC discourse. CMC was perceived less as a faster and more effective means of communication and exchange of messages, but more as a public sphere where different views (of the University funding crises) and values were confronted, where meanings and collective understanding were created. The methods we applied thus far were not sufficient to make sense of what was going on via CMC and to gain deeper understanding of the process. We needed a map or guide to the new social space that was in the process of making via CMC.

In this moment the significance of a social theory informing our inquiry emerged. I found in Habermas' theory of communicative action such a guide, enabling us to explore not only what linguistic expressions or acts (as Habermas calls them) *mean* but also what they *do* and what they *produce* in the life-world of participants.[3] Bringing Habermas' theory into the scenario of the empirical inquiry put new requirements on the interpretation of e-mail postings. In this context the purpose of hermeneutic interpretation became to develop understanding of communicative practices in CMC and reveal hidden forms of distorted communication that impacted upon the lifeworld of participants. Consequently, it was also intertwined with changing the research question or, more precisely, learning to ask more specific questions. For instance, some of the questions were: how participants used language to express themselves in e-mail messages and how such e-mail *linguistic acts* produced a particular type of action; in what ways such communicative practices (the linguistic acts and social actions of participants) shaped discourse, how they framed perceptions and problems to be resolved, established personal and collective identities, legitimised power relations and the production of organisational knowledge; and finally, to what extent such communicative practices were enabled and assisted by the particular features of CMC used.

While our inquiry originally started as an ethnographic study (aiming to improve understanding of how CMC was appropriated in a particular social context and culture), it was transformed on the way due to both our improved understanding and the evolving nature of CMC as an extended social space. As we adopted the new map (a particular critical theory) to guide our examination of the CMC as a social space, we faced problems characteristic of any critical research inquiry. The relationship between the theory, research questions and the empirical methods surfaced as a key issue for our research strategy.

The main difference brought by Habermas' theory was the change of focus from interpretation of individual e-mail discussions (that is their meaning within the context of consultation and the University broader context) to interpretation of the communicative practices and dialogical structures embedded in the e-mail discussion. From the text of the flow of participants' e-mail discussions we aimed to make sense of their linguistic acts and actual actions, that is what they wanted to achieve and how. Furthermore, we aimed to reveal how their linguistic acts and social actions affect their everyday lives, shape their beliefs, (re)create their identities and power relations, as well as legitimate knowledge (what is true, good, right). The underlying question here is to what extent specific communicative practices and resulting symbolic reproduction of the lifeworld are assisted or discouraged by CMC. To be able to achieve research objectives, to reveal the complex texture of social actions, system and social integration we adopted *critical ethnography* (Thomas, 1993; Myers and Young, 1997; Myers, 1997; Forrester, 1992). We adopted critical ethnography to help us discover how an apparently open and participative CMC discussion was in effect undemocratic; how CMC, introduced as a means for accessibility of information and transparency of the process, in fact enabled distorted communication, disempowerment of participants, and preservation of the existing power structure.

I would like to note here that we, as researchers, were very much aware of the uncertainty and fragility of our interpretations. As it is well argued by critical hermeneutics (the underlying philosophy of critical ethnography) "interpretations will never be linguistically unproblematic" (Kincheloe and McLaren, 2000, p. 289). For this reason our interviews with both participants and non-participants in the CMC discussion had the form of a dialogue. We shared with them interpretations and explanations of critical issues trying to achieve mutual understanding. This aspect of transition from interpretive ethnography to critical ethnography is nicely expressed by a quote:

"The paradigm is no longer the observation but the dialogue— thus, a communication in which the understanding subject must invest a part of his subjectivity, ... in order to be able to meet confronting subjects at all on the intersubjective level which makes understanding possible" (Habermas, 1973, p. 10).

Relationship Between a Theory and Methodological Strategies

The focus of this section is exploration of linkages between a theory and methodological strategies. The discussion reflects on the use of Habermas' theory of communicative action (1984, 1987) and methodological strategies applied in the research of the University CMC discourse. For that purpose I analyse two moments or substantive issues that exemplify the nature of this relationship: a) interpretive analysis of linguistic acts and social actions in CMC discourse, b) interpretation of relation between social interaction, system integration and social integration. To make it more comprehensible to the reader I illustrated graphically my interpretation of this part of Habermas' theory (in Figure 1).

Linguistic acts and social actions in CMC discourse

The speech or linguistic acts are observable parts of linguistically mediated social interaction. Unlike face-to-face interactions that also involve body language, in CMC a researcher (as well as a participant) experiences only textual expressions (e-mail messages) from which one makes sense of linguistic acts, actors' intention and social actions. Following Habermas (1984), the CMC discussion can be analysed at two levels of social interactions (presented in Figure 1):

- the level of linguistic acts and

Figure 1. The Framework for Critical IS Research Inquiry Informed by Habermas' Theory of Communicative Actions

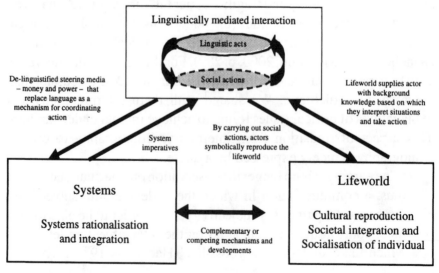

- the level of social actions constituted by individual linguistic acts.

To understand the performative aspect of linguistic acts (what words actually *do*), we (the research team) had to comprehend social actions accomplished by these acts. First, we had to find out what were the actual goals and intentions of participants. Second, we had to discover their orientation in pursuing the goals. During the consultative process, the President was pretty explicit, in his e-mails and in the *Restructuring* document, that his objective was to solve the financial crisis by restructuring the University (including a centralised model for staff funding). He publicly presented his views and his particular solutions via CMC and invited staff to respond to them. However, despite significant criticism he himself did not engage in debate about contested issues. The way the President engaged in the argumentation process suggests that he adopted a success-oriented attitude and thus his action may be interpreted as *strategic*. However, the analysis of his linguistic acts indicates that he did not want to show his strategic intent and that he upheld the appearance of communicative action (acting as if he is oriented to mutual understanding). In Habermas' theory this action is called a covert strategic action.

The attitudes of staff participants in the consultative process can be classified in two broad groups: one group that responded to the President's invitation for debate by acting strategically themselves and the other that attempted to act communicatively. By accepting the President's invitation to be consulted and by adopting CMC productively to engage strategically so as to counteract his strategic action, staff from the first group established their identity as more or less successful players in the game and relevant negotiators regarding conflicting matters. Other staff participants believed that the President's invitation to consult with staff meant that he wanted to establish mutual understanding with staff and define problems cooperatively. They understood the consultative process and the public debate via CMC as an opportunity for a *community dialogue* and a communicatively achieved agreement (they would typically start their message with *Dear Colleagues*, indicating that they are talking to all staff, not only to the President). These participants undertook a communicative action trusting that the President did too.

For the President the use of CMC was essential to undertake covert strategic action:

- He used CMC (e-mail especially) to expose his ideas and proposals to public scrutiny and criticism, thus establishing an appearance of an open dialogue, free criticism, unrestricted debate in which *everybody can have their say.*

- He interacted with staff via CMC on many occasions, attempting to show his sincere intention to listen and establish trust, but never engaged in an argumentation process.
- The e-mail discussion created such a huge number of different comments, ideas and proposals that, without careful analysis, it was not possible to make sense of what the University community wanted.

In such a way he used CMC instrumentally to conceal his strategic intent and pretend to act communicatively. This partially explains why he took the trouble of conducting the consultative process (first contradiction).

Staff, on the other hand, used CMC to achieve their goals as well. CMC enabled their voices to be heard. However, those who behaved strategically used CMC to influence and persuade the President. Others who behaved communicatively perceived CMC as a forum for achieving mutual understanding about the funding problems of the University and coordinating actions of all involved.

The above analysis illustrates the methodological strategy adopted at a micro level of social interactions. The analysis of linguistic acts and social actions went back-and-forth: first, the meaning of linguistic acts was derived from the text, within a limited understanding of the context (preceding flow of e-mails, other parts of the consultative process, history of the University, etc.); second, the actions were interpreted based on the meaning of linguistic acts; third, linguistic acts were reinterpreted as constituents of social actions.

This analysis, as a particular kind of hermeneutic circle that never ends, is a micro analysis of social interaction that reveals characteristics of communicative practices (e.g., strategic vs communicative acting, covert strategic actions) and provides a limited understanding of this process. To move to a new level of understanding, we had to broaden our investigation to consider the implications on both economic viability of the University (*system* aspect) and the lifeword of participants. In other words our methodological strategy had to change to embrace macro analysis as well (that is, include all three components from Figure 1.).

Systems integration versus lifeworld integration

By performing linguistic acts and carrying out social actions, participants in the CMC debate not only pursued their goals, they also defined a situation and a problem at hand, they presented themselves and recreated personal and group identities, they (re)established their position and legitimacy, they maintained or altered their working relationships, etc. Linguistic acts and social actions cannot be fully understood without their impact on *systemic integration* and *social integration* (Habermas, 1987). Namely, interpreting Habermas in the organisational context, the University can be seen simultaneously as the system and as the lifeworld of its members (Figure 1). A system aspect of the University involves its material and intellectual production, its economic foundation, administrative and management structure, rules and regulations, and the like. It is largely determined by Government regulations, policies and funding. In the CMC discussion it is pointed out that in order to survive, the University needs to consider economical, efficient, and effective delivery of courses, needs to have flexible staff management, to be cost effective, etc. In a word, it was driven by *purposive rationality*.

The lifeworld, on the other hand, is the symbolically (re)created, taken-for-granted universe of daily social activities of members. It consists of unproblematic, cultural knowledge shared by University members, that involves vast and unexpressed sets of beliefs, convictions, tacit assumptions, values that are in the background of social interaction. Members draw upon this knowledge to make sense of a situation, other actors, and their linguistic acts, and to take actions. When, for instance, participants in the University discussion refer to *community*, identify themselves as *we*, they in fact have in mind their lifeworld. By acting communicatively and coordinating their actions based on mutual understanding, actors rely on membership in the University as a social group and their lifeworld, and strengthen the social integration of the group. In Habermas's words communicative actions serve as a medium for symbolic reproduction of the lifeworld.

To transcend the limitations of the micro analysis of social interaction, we adopted a methodological strategy to consider mutual relationships between social interaction, system integration and lifeworld integration (see Figure 1). By doing so we situated hermeneutical interpretation in a larger whole (the whole University and its environment).   Differentiation of actions oriented to success and actions oriented to reaching understanding enables distinction between systemic integration and social integration. System integration operates through strategic action, driven by economic and administrative rationality, and disconnected from values and norms of members. On the other hand, the lifeworld is reproduced and rationalised by way of communicative action, in which intersubjective understanding, achieved through language, is a basis of action coordination. These are mechanisms by which social integration is achieved.

In the case of the University X systems concerns—as expressed by the President—include effectiveness of its operations, efficient and flexible academic structure, centralised resource allocation, ability to earn income, etc. The key action in the President's Restructure document was centralisation of staff funding presented as an imperative for the University. Interestingly he did not use his legitimate right to make such a decision without any consultation. He chose to pretend that he was acting communicatively thus heeding lifeworld concerns of community while in fact he acted strategically (concerned with system issues). He presented his solution as resulting through the common will anchored in the communicative practices that, to quote his words, *"emerged from collegial processes traditional in a university, as reflected in institutional discourse and related consultative activities"*. From the interpretation of (intended) impacts on the lifeworld, we were able to understand why he did so. By concealing his strategic action and presenting it as a communicative action, he aimed to legitimate it in the community and gain broad acceptance by the University members. From his point of view, this was necessary for effective implementation of his restructuring proposal. This line of interpretation from social interaction complex to lifeworld, or to systems issues and back, helped us reach a fairly sufficient and convincing explanation of the first contradiction.

As to the second contradiction, that is, how could an open, transparent and seemingly democratic process be used to advance a repressive outcome, it is necessary to view the use of CMC in the consultative process from the macro perspective of the interplay between systems and lifeworld.

In order to answer this question, we focused on the argumentation process, the ways claims and arguments were raised, responded to or ignored. A remarkable point was that the President never responded to the critique and counter-arguments raised by staff. While he declared that he was open to any suggestion he never engaged in a discussion about substantive issues. Moreover, staff rarely paid attention to, supported or rejected other staff's criticism, arguments and counter-arguments. The CMC discussion did not have a dialogical structure. The intensive traffic of e-mails, however, preserved the illusion of a dialogue. We interpreted it as a hidden mechanism reinforcing the existing power structure. Furthermore, we found out (from the interviews) that as a result the staff felt that he pushed the lifeworld concerns and systemic imperatives further apart. Eventually, staff became aware that *system imperatives* were given preference to community concerns and social integration. The conduct of the consultative process via CMC made it possible for the President to exert strategic influence on the decisions of other participants "while bypassing processes of consensus formation in language". As a result the lifeworld context of participants "gets *devalued*: the lifeworld is no longer necessary for coordinating actions" (Habermas, 1987, p. 281).

This is precisely what Habermas perceives as one of the dangers of increased complexity of modern organisations and society: when systems integration takes over and subsumes social integration, this leads to 'colonization' and erosion of lifeworld. The interpretation of the CMC discussion in the light of the two competing principles of societal integration and system rationalisation in the University re-structuring process enabled us to suggest a possible explanation of how an open and transparent CMC discussion was used to advance a repressive outcome (the second contradiction).

I have to note here that this example illustrates only one possible constellation among a range of possible constellations and interdependencies between systems and lifeworld. Systemic and societal integration are not necessarily competing developments either. They can in fact be complementary. On one hand, systems maintenance and development can be subject to the substantive and normative restrictions of the lifeworld. Conversely, societal integration through communicative action can be subject to the constraints of material reproduction. However, only if a rationalised lifeworld of a social group subjects the imperatives of system maintenance to the needs of its members, could an organisation hope to become emancipated (Wellmer, 1994). Based on the research outcomes from this inquiry, I can foresee the need for a new methodological strategy to go further and explore necessary organisational conditions and requirements for technological support to assist members in their struggle towards an emancipated organisation.

The analysis in this section illustrates methodological choices made in linking a particular critical theory, research methods and emerging research questions. For instance, the reflection on the methodological strategies in this chapter illustrates how Habermas' theory is injected into hermeneutical circles to guide interpretation of social and cultural texts produced via CMC, in order to connect the microdynamics of everyday social interaction with macro-dynamics of social structures.

CONCLUSION

The question of methodology in critical theory-informed IS research cannot be reduced to the 'problem' of the lack of empirical methods. Methodology, understood as an overall strategy of conceptualising and conducting research, is concerned with choices about linking critical theory, empirical methods and research questions in specific IS research situations. First, a choice of a critical theory sets a research agenda and poses specific research questions. Second, based on the epistemological assumptions and the kind of questions investigated, appropriate methods and their application need to be considered. Third, methodology is concerned with principles and processes of constructing scientific knowledge and making changes in social IS practice. Methodological debate, therefore, needs to address a much broader range of issues beyond the narrow view of specific critical empirical methods.

While the uniqueness of critical theory's methodology has been associated with reflexivity and dialectic character, it has not been associated with specific research methods. However, some empirical research methods are more likely to be appropriate than others. For instance, critical ethnography and participatory action research seem to share a common ground with critical epistemology and therefore are more likely to be appropriate for critical qualitative research. An interesting question for future debate may then be under what conditions certain empirical methods can be applied in critical theory-informed IS qualitative research. Another possible direction would be to develop more specific critical research methods either within a single research program (informed by a particular critical theory) or across programs.

Given a considerable diversity of contemporary critical research programs and the novelty of critical research in IS, comparisons of methodological strategies would be extremely interesting. This may, for instance, include comparisons of two critical research programs similar to the one by Morrow and Brown (1994) who compared a program informed by Habermas' theory of communicative action with a program informed by Giddens' theory of stracturation.

Critical research methodology has yet to reach its potential in qualitative IS research. For IS researchers, to engage in critical research is to participate in critical thinking about IS and organisations, guided by a vision of organisations as communities and of its members as emancipated workers. It is, to use Kincheloe and McLaren's words, *a pragmatic's hope in an age of cynical reason.*

REFERENCES

Bernstein, R.J. (Ed.) (1994). *Habermas and modernity.* Cambridge, Mass: MIT Press.

Cecez-Kecmanovic, D. (1987). *Social Information Systems: Concept, models and technologies.* Sarajevo: Savjet DSI SRBIH (in Serbocroatian).

Cecez-Kecmanovic, D., Treleaven, L., and Moodie, D. (2000). CMC and the question of democratisation: A University field study. Hawaii International Conference on Systems Sciences HICSS'2000. Hawaii, 3-8 January.

Cecez-Kecmanovic, D., Moodie, D., Busuttil, A. and Plesman, F. (1999). Organisational change mediated by e-mail and intranet – An ethnographic study. *Information Technology and People*, 12(1), 9-26.

Fay, B. (1987). *Critical social science: Liberation and its limits.* Ithaca, NY: Cornell University Press.

Forrester, J. (1992). Critical ethnography: On fieldwork in Habermasian way. In Alvesson, M.and Willmott H. (Eds.). *Critical management studies.* (pp. 46-65). London: Sage Publications.

Garrison, J. (1996). A Deweyan theory of democratic listening. *Educational Theory*, 46, 429-451.

Habermas, J. (1973). *Theory and practice.* Boston, MA: Beacon Press.

Habermas, J. (1984). *The theory of communicative action – Reason and the rationalisation of society.* (Vol I), Boston, MA: Beacon Press.

Habermas, J. (1987). *The theory of communicative action – The critique of functionalist reason.* (Vol II), Boston, MA: Beacon Press.

Held, D. (1980). *Introduction to critical theory.* Berkeley and Los Angeles, CA: University of California Press.

Hirschheim, R. and Klein, H. (1989). Four paradigms in information systems development. *Comm. ACM*, 32, 10, 1199-1216.

Hirschheim, R., Klein, H. and Lyytinen, L. (1996). Exploring the intellectual structures of information systems development: A social action theoretic analysis. *Accounting., Management. and Information Technology*, 6,1/2, 1-64.

Iivari, J., Hirschheim, R. and Klein, H. (1998). A paradigmatic analysis contrasting information systems development approaches and methodologies. *Information Systems Research*, 9, 2, 164-193.

Kiesler, S. and Sproull, L. (1992). Group decision making and communication technology. *Organisational Behaviour and Human Decision Making Processes*, 52, 96-123.

Kincheloe, L.J. and McLaren, P. (2000). Rethinking critical theory and qualitative research. In Denzin N.K. and. Linkoln Y.D (Eds.). *Handbook of qualitative research.* (2. edition), (pp. 279-313). Sage Publications.

Klein, H.K. (1999). Knowledge and methods in IS research: From beginnings to the future. In Ngwenyama, O., Introna, L., Myers, M.D. and DeGross, J.I. (Eds.). *New Information Technologies in organizational processes—Field studies and theoretical reflections on the future of work.* (pp.13-25). IFIP, Boston: Kluwer Academic Publishers.

Lyytinen, K. (1992). Information systems and critical theory. In Alvesson, M. and Willmott, H. (Eds.). *Critical management studies.* (pp. 159-180). London: Sage Publications.

Lyytinen, K. and Klein, H. (1985). The critical theory of Jurgen Habermas as a basis for a theory of information systems. In Mumford, E., Hirschheim, R., Fitzgerald, G. and Wood-Harper, T. (Eds.). *Research methods in information systems.* (pp. 219-236). Amsterdam: Elsevier Science Publishers (North-Holland).

Myers, M.D. (1997). Critical ethnography in information systems. In Lee, A.S., Liebenau, J. and DeGross, J.I. (Eds.). *Information Systems and qualitative research.* (pp.277-300). Chapman and Hall.

Myers, M.D and Young, L.W. (1997). Hidden agendas, power and managerial assumptions in information systems development – An ethnographic study. *Information Technology and People*. 10, 3, 224-240.

Morrow, R.A. and Brown, D.D. (1994). *Critical theory and methodology*. London: Sage Publications.

Schwandt, T.A. (2000). Three epistemological stances for qualitative inquiry: interpretivism, hermeneutics, and social constructivism. In Denzin N.K. and. Linkoln Y.D (Eds.). *Handbook of qualitative research* (2ed.). (pp 189-214). London: Sage Publications.

Siegel, J. Dubrovsky, V., Kiesler, S. and McGuire, T.W. (1986). Group processes in computer-mediated communication. *Organizational Behaviour and Human Decision*, 37, 157-187.

Sproull, L. and Kiesler, S. (Eds.). (1991). *Connections: New ways of working in the network*. Cambridge, MA: MIT Press.

Thomas, J. (1993). *Doing critical ethnography*. Newbury Park, CA: Sage Publications.

Treleaven, L., Cecez-Kecmanovic, D., and Moodie, D. (1999). Generating a consultative discourse: A decade of communication change. *The Australian Journal of Communication*, 26, 3, 67-82.

Walsham, G. (1993). *Interpreting Information Systems in organisations*. Chichester: John Wiley and Sons.

Wellmer, A. (1994). Reason, utopia, and the dialectic of enlightenment. In Bernstein J. R. (Ed.). *Habermas and modernity*. (pp. 35-66). Cambridge, MA: The MIT Press.

Wilson, F.A. (1997). The truth is out there: The search for emancipatory principles in Information Systems Design. *Information Technology and People*, 10, 3, 187-204.

ACKNOWLEDGMENT

In writing this chapter I benefited from discussions with my colleague, Cate Jerram, and from critical comments by two anonymous reviewers and Eileen Trauth, the editor of this book. To all of them I am deeply grateful. In this chapter I used as an example the research on CMC in a University consultative process in which my colleagues, Lesley Treleaven, Debra Moodie, Andy Busuttil and Fiona Plesman took part in various stages, each contributing in her or his own way.

ENDNOTES

1 The University staff involved approximately 250 academic and 420 general staff.

2 Richard Bernstein, the editor of *Habermas and Modernity* (1994) dedicated the book to his "Yugoslav *Praxis* colleagues who have been so courageous in fostering the ideals of democratic socialism by their words and deeds" (p. 32). In the Introduction he paid tribute to philosophers and social scientists of the Yugoslav *Praxis* group who strongly opposed Stalinist tendencies in Eastern Europe and advocated "the principles of self-management and participative democracy at all levels of society" (p. 32). The Yugoslav journal *Praxis* and the Summer school on the island of Korcula "became a meeting ground for progressive left intellectuals and students from Eastern and Western Europe as well as from English- speaking countries" (p. 31), including Bloch, Marcuse and Habermas.

3 For me, the choice of Habermas' theory of communicative action was natural as the assumptions and values the theory is based on coincide strongly with my own. However, this was not necessarily the case with other team members who did not have prior knowledge of the theory. Nevertheless, this did not obstruct our collaborative interpretation, as they were not biased either in favour or against this theory.

Chapter VII

Analysis by Long Walk: Some Approaches to the Synthesis of Multiple Sources of Evidence

Steve Sawyer
Pennsylvania State University, USA

INTRODUCTION

Through this chapter I make two contributions. First, I provide both conceptual guidance and practical advice for information systems (IS) scholars who are involved in multi-method research, with a particular focus on conducting multi-method analysis. Second, and as a means to achieve the first contribution, I detail some of the principal components of multi-method research. Multi-method research is based on the premise that analysis of separate and dissimilar data sets drawn on the same phenomena will provide a richer picture of the events and/ or issues than will any single method. While valued by many IS scholars, multi-method-based research to study the roles of information and communication technologies (ICT) in social organization is under-explored as a set of coherent techniques. In response I put forth a set of observations that arise from my own multi-method research experiences (see Guinan, Cooprider and Sawyer, 1997; Sawyer, Farber and Spillers, 1997; Sawyer, 2000b; Crowston, Sawyer and Wigand, 2001).

For this chapter multi-method means a combination of data-collection approaches, such as survey collection and field work, drawn on the same phenomena (Sawyer, 2000b; Brewer and Hunter, 1989). By analysis I mean the process of discerning findings from data. As context for this discussion I draw on my ongoing research into organizational computing infrastructure changes and enterprise resource package (ERP) installations (see Sawyer, forthcoming, 2000a, 2000b; Sawyer and Gibbons, 2000; Sawyer and Southwick, 1996, 1997). My discussion on multi-method research reflects the idiosyncratic blend of concepts, personal preferences and contextual circumstances through which an interpretive researcher sees the world. However, many of the issues I raise may be equally viable for IS scholars with different epistemologies than mine.

A multi-method approach to research on the uses of ICTs involves several data-collection techniques organized to provide multiple but dissimilar data sets regarding the same phenomena. By dissimilar I mean that they include different forms of data. For example, using participant observation and laboratory experiments is one way to conduct multi-method research (see Sproull and Kiesler, 1991). The observational data is typically textual and open ended, relatively unstructured and context dependent. Data derived from the experiments is typically de-contextualized, numerical and highly structured.

Multi-method research is typically done by drawing on data-collection methods that accommodate each other's limitations (Jick, 1979; Gable, 1994; Gallivan, 1997). For example, Sproull and Kiesler (1991) used the observational data to provide insight into the context and the experimental data to provide insight into observed behaviors. Further, both the conceptual bases and data collection techniques help to shape the phenomena of interest. So, there are many ways to conduct multi-method research (Brewer and Hunter, 1989). Here I focus on "multi-method fieldwork": blending fieldwork with surveys, as this is what I most often do in my research. My conceptual bases are rooted in social theory (see Sica, 1998), and this combination of methods and theory leads me towards different questions than would someone who draws on psychology and combines fieldwork with experiments (see Sproull and Kiesler, 1991).

Fieldwork includes participant observation, interviewing, and collecting archival records—characteristics of both intensive and prolonged involvement with the social units being studied. (Jackson, 1987). Surveys involve data-collection instruments (often self-administered) to collect responses to *a priori* formalized questions on predetermined topics from a valid sample of members of identified social categories. Surveys are a mainstay of the quasi-experimental field research tradition on which most IS research is based; survey-based studies comprise more than 49% of research done on the use of ICT in organizations (Orlikowski and Baroudi, 1991). Explicit multi-method studies represent about 3% of the same research base (Gallivan, 1997).

A multi-method fieldwork approach implies that surveys are used in a manner that differs from the more common quasi-experimental field research use. In the latter, survey data are extracted from the field and quantified. Non-survey data are used to support or enrich findings from survey data. A multi-method approach sees the two forms of data collection as co-equals. That is, each data-collection method must both stand on its own (independence) and also be combine-able (interdependence) (Kaplan and Duchon, 1988; Gallivan, 1997; Jick, 1979; Brewer and Hunter, 1989). I discuss these concepts in more detail in section three.

This process of combining multiple data sets is often called triangulation (see Jick, 1979). Triangulation is the analytic act of identifying similar findings from different data sets. Essentially, this suggests seeing the same "research event" from different perspectives. This strict definition of triangulation reflects a positivistic perspective that there is a single truth to be observed (Falconer and Mackay, 1999a, 1999b, 2000). Such a strict definition also implies that a common research perspective is embodied in a triangulating analysis (Jones, 1999; Falconer and Mackay, 1999a). In this chapter I do not engage directly in the emerging debate about either the viability of the commonly held view of triangulation or the more broadly philosophical debates on paradigmatic pluralism (see Jones, 1999). Instead of focusing on triangulation's meaning(s), later in this chapter I discuss the analysis of combined data sets in terms of comparison and contrast analyses.

On a more pragmatic level there are few common conventions, and even fewer analytic techniques, to describe the process of the analysis of combined data sets (one exception is the positivist's multi-trait/

multi-method matrix described by Campbell and Stanley, 1966) (Gable, 1994; Howe and Eisenhardt, 1990; Williams, 1986; Jick, 1979). This often limits the value of this type of research to the broader community because describing the methods used in doing such a multi-data set analysis is both important for establishing credibility and space-consuming (Seidler, 1973; Lincoln, 1995; Sutton and Staw, 1995). The need to write extensively about non-standard methods typically comes at the cost of reducing the space devoted to discussing findings.

Both fieldwork and the more common survey-based approaches have strong ideological bases. A multi-method approach that combines surveys with fieldwork seeks to integrate (or at least bridge) these perspectives. For example, Langley (1999) argues that "synthetic" methods — where various forms of data are linked together in emergent analyses, represents an under-explored area of methodological development. The result of such a linkage is a new method, not just an aggregation of existing techniques (Brewer and Hunter, 1989, p. 17). For example, in Sawyer (2000b) I contrast the multi-method research done by Kaplan and Duchon (1988) and Guinan, Cooprider and Sawyer, (1997). By making this contrast, I illustrate the various ways these studies draw on the strengths of the combination of data-collection methods (Jick, 1979; Gallivan, 1997).

The chapter continues in four sections. In the first section I provide an overview of Mid-Sized University's (MSU) computing infrastructure change as the context for the discussion of conducting multi-method research. Section two contains a discussion of the elements that underlie a multi-method research effort. In section three I present a set of issues that arise in the conduct of multi-method analysis and in the final section I highlight some unresolved aspects of multi-method approaches to research on ICTs uses.

AN EXAMPLE: MSU'S COMPUTING INFRASTRUCTURE CHANGE

The context for our discussion of multi-method analysis is an ongoing study of the computing changes at one organization. Since early 1992 MSU has been engaged in installing both a client/server-based computing infrastructure and an enterprise resource package (ERP) suite to replace their mainframe computing infrastructure and proprietary, stand-alone administrative information systems. The research goal has been to identify how changes to MSU's computing infrastructure are manifested in the technical, social, and administrative structures of the organization.

MSU is a private, research-oriented university which enjoys high name recognition, nationally and internationally. MSU's administrative and organizational structures are representative of typical U.S. universities of nearly 18,000 students and 4,000 employees. However, by 1993, three environmental factors constraining MSU's computing infrastructure created a situation demanding senior management's attention. These were: (1) an increasing workload required of MSU's mainframe systems, (2) a restrictive reliance on MSU's outdated legacy systems, and (3) an increasingly unmanageable tangle of administrative and academic computing networks, characterized by overlapping links and disparate technologies.

These issues are typical of most academic computing systems (Alpert, 1985; El-Khawis, 1995; McClure and Lopata, 1996) and many mid-sized and large organizations. Facing this scenario, MSU's CIO made the decision to revamp the computing infrastructure to take advantage of new client/server technologies and purchased software (the ERP and some additional software packages).

The MSU case is similar to many other organizations who are changing their computing infrastructure and installing ERP software. By ERP I mean an integrated suite of software modules that partially automate an organizations key processes (such as manufacturing, accounting, marketing, HR, payroll, etc). Relying on an ERP is typically a major change to an organization's operations (Davenport, 2000). Since these systems are vendor provided, they are not fully customized for any user. Thus, acquiring an ERP system requires extensive tailoring of both the package's functions and the organization's work practices (Sawyer, forthcoming).

The MSU research was designed as a longitudinal, multi-method effort spanning individual and organizational levels of analysis. At the individual level, the focus is on how the ERP systems, and the consequent work changes, are being interpreted. At the organizational level, the focus of research has been directed toward understanding how these systems affect the web of computing already in place at MSU. The research approach has also been designed to allow us to focus on the potentially differing perspectives towards ERP among individual users, technologists, and vendors. Five people have participated in data collection at various points while one has been part of the project from its inception.

Data-collection activities encompass participation with, and observations of, committees formed to work on specific aspects of the ERP initiative; interviews with managers and workers (both IS and line); document/records collection and surveys. The field work aspects include extensive observation, interviewing and archival record collection. Survey data includes two sets of employee assessments (focused on training and quality of work life). These surveys' data were provided to the research staff. Interviews varied by level of structure, with most being semi-structured and open-ended. The field notes record data collected from unobtrusive observations, from participation in committees and meetings, and from informal social interactions. There are two types of field notes for each observation, interview, or interaction. The first type is a chronology of events and actions; the second is a more free-flowing account of perceptions, stories, and anecdotes. The chronology serves as a record of observations. The account serves as a record of the observer's perceptions.

Survey data were gathered by the human resource department and were designed to provide insight into changes in the job classification structure, to plan for IT-focused training, and to gauge employee readiness for the computing changes. The two surveys contain 40 questions (plus some demographic requests) with most of them using either five-point Likert scales or yes/no responses. The survey data were not collected in support of an explicit predictive model.

ELEMENTS OF A MULTI-METHOD RESEARCH APPROACH

Like many forms of research, multi-method approaches are guided by a number of factors such as validity and generalizability. Factors common to most research approaches are well-documented in other work and not discussed here (see, for example, Brinberg and McGrath, 1984; Creswell, 1994; Danzin, 1970). Of more interest for this chapter are the factors specific to multi-method research such as: the roles of theory, method independence, insulation, data interdependence, analytic integration, and data comparability versus contrast (Brewer and Hunter, 1989). In the rest of this section, I outline these factors and explain how the MSU research was designed with these in mind. The factors are listed and defined in Table 1.

The Roles of Theory

Central to most research is the development and/or testing of theory (Blalock, 1971; Popper, 1968; Weick, 1995). Contributions to theory can be seen along a continuum from development to testing (Sutton and Staw, 1995; Glaser and Strauss, 1967; Yin, 1989; Blalock, 1971; Bagozzi, 1979; Hoyle, 1995).   However, theory development and/or testing is often not done well in organizational research, a subset of which is research on organizational computing infrastructures (Sutton and Staw, 1995; Weick, 1995; Merton, 1967). For example, Grunow (1995) reports on the methodological and theoretic approaches to 303 papers concerning organizational research. He found that 78% of these papers did not align theory to their research questions and 82% of the papers did not contribute meaningful results to theory development. As Sutton and Staw (1995, p. 371) state: "... references, data, variables, diagrams and hypotheses are not theory."

Even if Sutton and Staw (1995) are correct and Grunow's (1995) analysis is accurate, some form of theoretical rationale still forms the basis of the analysis and discussion sections of most scholarly papers on ICT uses in organizations. One benefit, and a primary differentiator, of multi-method research is multiple data sets. This suggests two roles for theory. First, theory can be a source of guidance on how to develop multiple data sets by helping the researcher focus on the types of data needed. Second, theory is also the means for uniting the various data-

Table 1: Six Issues with Multi-Method Data Collection

Role of theory	Used to guide the study and establish relationships between multiple data sets
Independence	The effect of one data-collection method to another
Insulation	Exposing subjects to effects of multiple waves of data collection
Interdependence	Providing for intentional links between data
Integration	Combined analysis of multiple data sets
Comparability v. Contrast	Analysis highlights differences caused by different type of data and can lead to incongruities in analysis.

from Brewer and Hunter (1989)

collection approaches because it provides common concepts that help to structure the data-collection efforts. While some authors protest against using formal theory (see Van Maanen, 1995a; 1995b), theory serves as a stabilizer for multi-method-based research. A theory-based approach helps sort through the blur of reality, providing a way to characterize observation and interpretation (Weizenbaum, 1976). Vaughan (1992) calls this "theory elaboration" and Weick (1995, p. 385) calls it "theorizing." Hence, multi-method research approaches imply the use of *a priori* theories, though the actual linkages among these theories may not emerge until data collection/analysis.

For example three *a priori* theories form the interpretive frame of the MSU research. Structuration theory provides a way to relate social structures and physical structures (such as the ICTs being used) through the ongoing actions of the social actors (Giddens, 1984; Giddens and Turner, 1987; Orlikowski, 1992; Orlikowski and Robey, 1991; DeSanctis and Poole, 1994; Koppel, 1994). This perspective provides a lens for viewing organizational and departmental action. Work design provides a perspective extending from individual through the departmental level. This perspective focuses on work as a set of tasks that are related to group and social norms (Hackman,1977; Hackman and Oldham 1980; Nadler, 1963; Adler, 1986; Kelley, 1990). Punctuated equilibrium relates both across time (Gersick, 1988, 1989,1991; Miller and Friesen, 1980; Tushman and Romanelli, 1985; Romanelli and Tushman, 1994).

These theories were selected because they overlap at various levels of analysis, explicitly incorporate the temporal nature of the study, and together provide a means to structure data collection. This set of theories also provides a means to organize analysis of the survey data. The difficulty with relying on *post-hoc* theory is that there may be little linkage among the data sets and the study becomes less of multi-method and more of a series of studies on a common topic (Gallivan, 1997). In essence, this is the conceptual contention made by Falconer and MacKay (1999b). Kaplan and Duchon (1988) also highlight the pragmatics of trying to merge dissimilar data into a coherent set by using a form of grounded theory, pointing out how difficult it is to redirect the analysis effort in light of unexpected findings.

Independence

One value in using multiple collection methods is that they draw data from the same group of people. However, these people interact with the researcher(s) in several ways over the course of data collection (as participants in interviews, as subjects of observation and as respondents to surveys). This means that the survey effort is also related to the interview effort in that they are typically done by the same researcher(s). Hence, multi-method researchers must consider how their *total* presence will affect subjects.

To address independence the MSU research was planned in multiple phases and focused at multiple levels. This allows for data collection to be separated by level of analysis and provides for some control over selection of participants. Further, this phased approach allowed multiple researchers to have distinct roles in each phase. For example, I worked closely with several groups and was the only person to interact with senior IS and organizational leaders. Other researchers were given specific projects and/or groups of people to follow or interview. In this way we reduced the confusion over having multiple researchers, reduced potential conflicts of interest over the purpose of the research, mitigated concerns that the research staff would be seen as reporters for senior leaderhip, provided for some "territorial" clarity, and minimized confusion for the people at MSU. However, one limitation of this approach is that we had virtually no overlap in respondents/participants among the research team.

Insulation

Inherent to multi-method fieldwork is the disturbance caused by the (multiple) presence(s) of the researcher(s). Approaches to reducing this disturbance lie on a continuum from unobtrusive observation to direct participation (i.e., action research) (Argyris, Putnam & Smith, 1985). The surveyor leaves behind the anticipated and unintended effects that the questions in her instrument instigate. The fieldworker is an interpreter and contributor to both the events seen and roles played during her time in the field. Thus, the roles of the multi-method researcher remains a research decision open to interpretation.

The MSU research plan incorporates three ways to insulate data-collection efforts. First, the various levels of the organization are used as insulation. Each member of the research team is focused on particular aspects. And, while the formal and informal social networks of any organization suggest that there will be some awareness, using organizational level as a means of insulation is both easy (as levels exist) and flexible (since researchers can have different roles at different levels). For instance, one researcher at MSU became aligned with certain user groups. And, even though they later worked to follow a particular sub-project of the implementation, they maintained ties to the user group representatives included in that project. Second, the MSU research spaces data collection across time to provide an insulating effect. After a prolonged interaction with the IT unit (see Sawyer and Southwick, 1997), we returned on occasion to conduct follow-up interviews. Third, using unobtrusive and non-invasive data collection (such as archival record collection) helps to insulate one data set's effects on another. At MSU we have been included on a number of electronic mailing lists and listservs where our presence, while not hidden (as we make sure that other members are aware of our presence) is not invasive.

Interdependence

Balanced against the independence and insulation issues is the desire to create linkages between the various data sets. In the MSU study we stressed the value of interdependence over insulation. Following from this decision we approached linking data sets three ways. The most obvious way, as laid out above, is creating links via theory. This is why I've advocated that there must be some form of *a priori* theory that helps to structure the collection of data. A second means to create interdependence is to focus on overlapping concepts. Often these emerge from the interim analysis. For example, issues of time (an aspect of the *a priori* theory) led to interim analyses where the concept of temporal differences among various work groups involved in the implementation of the ERP arose. This led us to searching for discussions of temporal differences in other data sets (see Sawyer and Southwick, 1997).

A third form of linkage is to focus on events. This form of linking uses a particular event as a means to organize the data. In our analysis of the events related to one of the MSU sub-unit's installation of a client/server computing-based system we structured the analysis around different data sets by focusing on what was said, done, or related to particular aspects of the implementation process (see Sawyer and Gibbons, 2000). We have also begun to discuss analysis based on highlighting the connections among different sets driven by particular people or groups—a form of social network analysis.

Integration

Analyzing multiple data sets that are focused on a common phenomena often leads to paradoxical results. That is, data drawn from
different data sets may lead to contradictory findings (Sawyer, 2000b).
These contradictions represent the potential for new learning or for
exposing methodological flaws (Robey, 1995). Developing such
findings suggests the value of interim analyses to help define differences among data sets (Miles, 1979).

The need for ongoing analysis and the intense energy that field
work demands, the multi-method researcher must be both close and
distant to the data. To make sense of mixed forms of field data, ongoing
analysis is critical and deeply reflective analysis is demanded. Pre-
scriptive analytic techniques — available in more traditional experimental and quasi-experimental data analysis (e.g., Pedhauzer and
Schmelkin, 1991) —are not as well developed for qualitative analysis
(e.g., Miles and Huberman, 1994). In fact, that is one of the points that
gives rise to this chapter. Still, there is some guidance. For example,
using explanatory matrices — where issues form one axis, sources form
the other, and supporting data fill the intersecting cells — is one flexible
technique (Miles and Huberman, 1994). Another technique is to build
evidence chains, where an issue is stated and then the supporting
evidence is laid out. Both of these imply immersion in the data sets to
develop ways to categorize the corpus of data and to extract the relevant
segments. This guidance gives rise to additional operational issues
which we take up in the next section.

Is Analysis Comparison or Contrast?

This aspect of multi-method research focuses attention on the way
analysis is conducted. One approach is to focus on identifying the
overlaps between sets. As discussed above, this is the approach which
best aligns with attempts to triangulate findings across multiple data
sets (Jick, 1979; Jones, 1999). A contrast approach to analysis implies
that the multiple data sets allow the researcher to highlight and explore
incongruities and paradoxes that arise. Influenced by Robey's (1995)
conceptualizing the value of exploring paradox, in the MSU study we
have instead used contrast as the orienting approach (see Sawyer and
Southwick, 1997).

Relationship to Process and Variance Models

A multi-method fieldwork approach to research draws on data collected using techniques that often are tied to dissimilar epistemologies (Jones, 1999; Falconer and McKay, 1999a, 1999b, 2000). For example, most survey approaches to data collection are focused on developing factors or structures that explain or predict outcomes, and these are typically factor or variance models. Fieldwork data sets often imply a process or sequence theory (Abbott, 1995). Mohr (1982) argued that process and variance theories drew on different epistemological assumptions and should not be mixed. Within the IS literature, Markus and Robey (1988), and more recently, Jones (1999) have also argued for this delineation.

Shaw and Jarvenpaa (1997) go on to present a typology for combining process and variance approaches that has 18 forms, 16 of which are hybrid combinations. Each of these approaches allows for multi-method research. However, in each approach the means to incorporate process and variance leads to different issues. My intent in this chapter is to focus on analysis of multiple and dissimilar data sets, not to engage in the ongoing debates on process and variance models (see instead Mohr, 1982; Markus and Robey, 1988; Abbott, 1995; Langley, 1999; Weick, 1999). However, it is important to realize that developing multi-method research is likely to also mean confronting some epistemological choices and explicitly deciding on how to accommodate both process and variance models.

MULTI-METHOD ANALYSIS ISSUES

In this section I reflect on my efforts to conduct multi-method analysis, touching on some of the mechanical and philosophical issues along the way. My focus on multi-method analyses suggests that I draw on both fieldwork analysis and survey analysis. However, I do not spend our time on these topics. The literature and guidance on analyzing survey data is both extensive and well known (e.g., Miller, 1991; Pedhauzer and Schmelkin, 1991). The literature and guidance on fieldwork and other intensive and/or qualitative research approaches has a growing body of literature, and the companion chapters of this book contribute to this corpus (see also Miles and Huberman, 1994). My focus is on how to combine and/or move between multiple and dissimilar data sets drawn on the same phenomena—a form of synthesis. I would also note that most social scientists have done some form of this synthesis (such as using some pilot interviews to help structure a survey or archival data to assist in setting up an experiment). My orientation is towards the use of synthesis to allow me to move among data sets. The rest of this section contains a discussion of issues that arise in the analysis of multiple data sets.

Explanatory Matrices Provide a Means to Synthesize Dissimilar Data Sets

In the process of coordinating and tracking data collection the research team members have used vignettes and stories to help develop a shared understanding of what we are learning (e.g., Miles, 1979, 1990). The research design means that each of the research team members develops insights driven by their field work, use of survey data, and perspectives into the MSU transition. Through this planned – but informal – interim analysis effort several themes emerged. These themes are used to help frame a return to the field notes and organize data to support or refute their value. This framing was done using explanatory event matrices (Miles and Huberman, 1994; Miles, 1990). In an explanatory event matrix, the themes form one axis (typically the rows) and the sources of evidence form the other axis. The cells made from this matrix contain pointers back to the source of evidence relative to the theme (or concept) to which it relates (see Appendix 1 for more about explanatory matrices).

Two points follow from this use of explanatory event matrices. First, using these matrices creates evidence chains, an important element of any rigorous method. The chain of evidence allows other scholars to understand (if not replicate) the thinking that is embodied in the matrix. Second, explanatory event matrices are not tied to a particular method of analysis—they help to juxtapose themes and evidence. This means that themes drawn from different data sets can be set within the same matrix.

Non-Overlapping Findings Lead to Richer Insights

Kaplan and Duchon (1988) highlight that discrepancies between their field work and the survey data were instrumental both in nearly derailing the project and served as the source of the greatest insight. This contrasts with proponents of multi-method analysis who suggest that multiple data sets allow the researcher to draw the same conclusion by drawing on analyses the same phenomena from multiple perspectives (Jick, 1979). Focusing on where the themes and findings that emerge from analysis do not overlap has led to greater insight in my own research. For example, the findings of my dissertation (which primarily focused on analyzing survey data) suggested that software development teams did not benefit much from using an electronic meeting support (EMS) system (Sawyer, 1995). However, after more detailed examination of the 16 months of fieldwork data (and additional data collection), I was able to discern distinct differences in EMS use based on various software development team social practices (Sawyer, Farber and Spillers, 1997).

In the MSU study survey data suggested that most respondents were comfortable with the impending computing change. However, one of the most consistent observations was that the majority of the people who were to be affected by the ERP installation had high levels of uncertainty and were quite unsure of the changes to come. This difference highlights the value of the contra-indicating observation as an important signal for additional attention. These differences also highlight the value of multiple data sets – I have been able to draw on observations, conduct interviews and re-read documents to help re-examine evidence about how MSU's people are preparing themselves for ERP-enabled change.

Integrating Across Multiple Data Sets Is Time-Consuming

While the use of explanatory event matrices allows some juxtaposition of data sets, I find it difficult to be immersed in more than one set of data at a time. Given the large number of non-overlapping themes and findings in the different data sets, switching data sets means also switching from one set of findings and themes to another. Without some unified means to conceptualize these data sets, the range of findings and themes quickly overwhelms me. Kaplan and Duchon (1988) intimate this in their discussions of how the project team, with each member tied to a particular data sets' findings, had trouble understanding the findings and themes that arose from analysis of other data sets. In the context of the MSU data set, it was easy to dismiss the findings that arose from the survey data analysis. Given the investment in the fieldwork and the lack of involvement in designing and collecting the survey data, it was difficult to invest the time to understand this data set.

Moving between data sets is time-consuming: a personal "switching" cost. The Guinan, Cooprider and Sawyer (1997) paper reflects this in that we did not overtly include the findings that emerged from our field work, reporting instead a series of models derived from the survey data. However, these models were informed, and the analysis guided by, the field work. Thus, we minimized switching costs by privileging the survey data over the fieldwork data. I did this again in Sawyer and Guinan (1998): highlighting survey data findings but relying on (unreported) field work data to shape this analysis. In doing this I minimized the cost of switching data sets and perhaps also reduced the value of the potential insight. Gallivan (1997) notes that many multi-method studies are reported as distinct pieces. Moreover, I have taken the same path in my current study on MSU's computing infrastructure: focusing each paper on the themes that arise from a particular perspective and in doing so muting the other perspectives.

Privileging One Perspective on the Data

Implied in the discussion of switching between data sets is the importance of knowing one's data. Multi-method research means there are two (or more) sets of data drawn using different ontological and methodological approaches. This demands great intellectual flexibility (or perhaps naivete) for the researchers (Jones, 1999). In many cases, multi-method research approaches rely on multiple researchers, each tied to a particular method (as in Kaplan and Duchon, 1988). Other times, the research team members overtly privilege one perspective over the other (in order to reduce the difficulty of knowing deeply and the cost of switching between). In the MSU study, the six-plus years of study have added the additional complexity of data volume. With such a large corpus of evidence, it becomes difficult to organize and present the data in a way that allows one to comprehend the data set. In this way my experience with the MSU multi-method study mirrors some of the experiences Schultze (2000) writes on in her ethnography of information workers. Thus, difficulties in understanding large data sets are not particular to multi-method studies. However, the need to know dissimilar data sets further complicates the processes of understanding this data.

Using Interim Theories

The best way I have found to explore the data has been to nurture emerging hypotheses – to check out my intuition by returning to the data sets, pitting emerging theorizing against current data. Miles (1979) writes of this approach as interim analyses: carrying the results of a particular interim effort forward into future analyses. And, earlier, I wrote of explicitly incorporating this into the research method through the research team meetings. In the presence of multiple and dissimilar data sets, I've found this approach to be quite valuable. Coupled with using explanatory event matrices, this allows me to explore an emerging theory across several data sets (and in building the matrix I also develop my evidence chain as a record). In the cases where there are multiple researchers, this approach provides a basis for comparison and contrast. As an example, early in the MSU case we did a short study of a self-contained component to the larger project (see Sawyer and Gibbons, 2000). During the data collection, we (the two co-authors) developed different theories of how the installation of the new parking services computing system evolved. Both of us built matrices supporting our pet theories and then were able to negotiate through these to the point that the final analysis (represented as a case study) emerged.

Creating Linkages Between Data Sets: Taking the Long Walk

Relative to any mono-method analysis, multi-method data analysis takes added time to learn multiple and dissimilar data sets, to be able to switch between them, to identify and analyze the outliers, and to work through the myriad pet (or interim) theories that arise from within various data sets. This time is set against most scholars' personal agendas and perceptions of time. Such deep and consuming reflection must be set against the realities of academic work/reward structures (for granting, tenure and promotion), the topical movement that is a pertinent factor in contemporary studies of ICT use, and the current (journal-article-oriented) structures of IS scholarship.

Beyond the contextual observation, one under-explored issue relative to the synthesis of multiple data sets is data reduction — the process of abstracting themes from the evidence and then testing and modifying these themes across multiple data sets. In the absence of automated approaches (such as the use of factor analytic techniques in statistical analyses) to combine data sets, I have found reflection the best means. This reflection is often done via long walks which I have used to sort out a blur of facts and make some sense from so much. As biographies of many scientists (such as Einstein, Watson and Crick, Paling) have suggested, deep reflection is a common aspect of much research[2]. In essence, the long walk becomes a data-reduction technique. Perhaps the most discomforting aspect of this observation is that a reliance on individual skill in not replicable. This suggests that synthesis may be idiosyncratic.   This observation implies that the process of making the linkage is less important than is the ability to represent the link. In essence, this shifts the attention from method to finding. And, while this seems reasonable it also suggests the difficulty in developing a reasonable basis for how best to make these linkages.

DOING MULTI-METHOD RESEARCH: SOME CONCLUSIONS

One of the more persistent challenges in multi-method analysis is documenting the analysis of multiple data sets. This is one of the reasons I have repeatedly suggested developing explanatory event matrices. In practice, the matrix is often a simple table, with the cells containing pointers back to the proper sources. Less than one-half of the MSU field work is digital (much of it is pen and paper), so these matrices become shorthand structures for recall. I date them and often have multiple versions clipped together. Combined with reflective comments in the field notes (as suggested by Bogdan, 1972 and discussed above), this is my best way.

One common response to my plaint, above, is to use some form of computer-aided analysis software (such as NUD*IST or ATLAS-TI). However, I have found that they serve mainly to help maintain the structure of relations (embodied in my series of matrices with pointers). And, to use them it often requires either digitizing all the data or maintaining both a paper structure and a digital structure (in the software). Thus, I've moved away from using such tools. To me the difficult work is developing the structures and themes, not keeping track of them.

It also seems that most multi-method studies are based on teams of multiple researchers. Typically, each researcher had a particular method expertise (and perspective). For example, in the MSU study, one researcher has led the effort, working with various collaborators within the research framework. In this way each research team member has a pre-defined role. It may be that a multi-method approach demands multiple researchers, that the different data-collection approaches are tied to the particular researchers, and the integration across multiple data sets is, in practice, a negotiated arrangement among the research team. Certainly this is the way multiple methods were used by Kaplan and Duchon (1988).

Moreover, in each of the examples in this chapter, one data set (or at least one type of data) dominates the analysis. In effect, this is the way most survey-based research is done: the survey data set is dominant and all fieldwork (interviews, etc.) are explicitly or implicitly subjugated (Brewer and Hunter, 1989). The MSU approach reflects this, but the dominating data set is the corpus of field notes, making this more like a fieldwork study than I had ever anticipated. While sensible, privileging data sets suggest that any multi-method research approach evolves into a tiered research effort.

Conceptually, though, multi-method analysis is based on the premise that separate and dissimilar data sets drawn on the same phenomena will provide a richer picture of that event or issue than will a mono-method analysis. My experiences suggest that, while valued by social science and IS scholars (see Jick, 1979; Brewer and Hunter, 1989; Gallivan, 1997), multi-method research is still an under-developed approach. Still, even given its current level of ambiguity, multi-method approaches provide a robust frame from which we can work on developing better insights into, and more useful theories of, the inter-related roles of ICT's uses, and the formal and informal social organizations into which they are embedded.

REFERENCES

Abbot, A. (1995). Sequence Analysis: New Methods for Old Ideas, *Annual Review of Sociology, 21*:93–113.

Adler, P. (1995). Interdepartmental Interdependence and Coordination: The Case of Design/ Manufacturing Interface. *Organization Science 6(2)*:147 -167.

Adler, P. (1986). New Technologies, New Skills. *California Management Review (19)*1:9-28.

Alpert, D. (1985). Performance and paralysis: The organizational context of the American research university *Journal of Higher Education, 56(3)*: 242-281.

Argyris, C., Putnam, R., & Smith, D. (1985). *Action Science: Concepts, Methods, and Skills for Research and Intervention*. San Francisco: Jossey-Bass.

Bagozzi, R. (1979). The Role of Measurement in Theory Construction and Hypothesis Testing: Toward a Holistic Model. In *Conceptual and Theoretical Developments in Marketing*. Ferrell, O., Brown, S. and Lamb, C. (Eds.). Chicago: American Marketing Association.

Barley, S. (1990). Images of Imaging: Notes on Doing Longitudinal Field Work. *Organization Science*, *1*(3):220-249.

Benbassat, I. Goldstein, D. & Mead, M. (1987). The Case Research Strategy in Studies of Information Systems. *MIS Quarterly, 11*(3):369-386.

Blalock, H. (1971). *Causal Models in the Social Sciences*. Chicago: Aldine.

Bogdan, R. (1972). *Participant Observation in Organizational Settings*. Syracuse: Syracuse University Press.

Brewer, J. & Hunter, A. (1989). *Multi-Method Research A Synthesis of Styles*. Newbury Park, CA, Sage.

Brinberg, D. & McGrath, J. (1984). *Validity and Research Process*. Beverly Hills, CA, Sage.

Burkhardt, M. (1994). Social Interaction Effects Following a Technological Change: A Longitudinal Investigation. *Academy of Management Journal, 37*(4):869-898.

Campbell, D. & Stanley, J. (1966). *Experimental and Quasi-Experimental Designs for Research*. Chicago: Rand-McNally.

Collins, P. & King, D. (1988). Implications of Computer-Aided Design for Work and Performance. *The Journal of Applied Behavioral Science, 12*(2):173-190.

Creswell, J. (1994). *Research Design: Qualitative and Quantitative Approaches*. Thousand Oaks, CA, Sage.

Crowston, K., Sawyer, S. and Wigand, R., (2001). The Interplay Between Structure and Technology: Investigating the Roles of Information Technologies in the Residential Real Estate Industry, *Information Technology & People, 15*(2): in press.

Danzin, N. (1970). *The Research Act: A Theoretical Introduction to Sociological Methods*. Chicago: Aldine Publishing Company: 297-313.

Davenport, T. (2000). *Mission Critical: Realizing the Promise of Enterprise Systems*, Cambridge, MA: Harvard Business School

Press.

DeSanctis, G. & Poole, M. (1994). Capturing the Complexity in Advanced Technology Use: Adaptive Structuration Theory. *Organization Science, 5*(2):121-147.

El-Khawis, E. (1995). Campus trends: 1995, in Higher Education Panel Report, Number 85, American Council on Education, Washington DC..

Fairhurst, G. Green, S. & Courtright, J. (1995). Inertial Forces and the Implementation of a Socio-Technical Systems Approach: A Communication Study. *Organization Science, 6*(2):168-185.

Falconer, D.& Mackay, D. (1999a). Ontological problems of pluralist research methodologies, Proceedings of the Fifth Americas Conference on Information Systems, Milwaukee, USA.

Falconer, D. & Mackay, D. (1999b). The key to the mixed method dilemma, Proceedings of the Tenth Australasian Conference on Information Systems, Wellington, NZ.

Falconer, D. & Mackay, D. (2000). The myth of multiple methods Proceedings of the Sixth Americas Conference on Information Systems, Long Beach, USA.

Forrester Research, Sizing Commerce Software, Boston, MA, 1998.

Gable, G. (1994). Integrating Case Study and Survey Research Methods: An Example in Information Systems, *European Journal of Information Systems, 3*(2): 112-126.

Gallivan, M. (1997). Value in Triangulation: A Comparison of Two Approaches for Combining Quantitative and Qualitative Methods. in *Qualitative Method in Information Systems,* A. Lee, J. Liebenau, & J. De Gross (Eds), New York: Chapman & Hall:83-107.

Gersick, C. (1991). Revolutionary Change Theories: A Multilevel Exploration of the Punctuated Equilibrium Paradigm. *Academy of Management Review, 16*(1):10-36.

Gersick, C. (1989). Marking Time: Predictable Transition in Task Groups. *Academy of Management Journal, 32*:274-309.

Gersick, C. (1988). Time and Transition in Work Teams: Toward a New Model of Group Development. *Academy of Management Journal, 31*:9-41.

Giddens, A. (1984). *The Constitution of Society: Outline of the Theory of Structure.* Berkeley, CA: University of California Press.

Giddens, A. & Turner, J. (1987). *Social Theory Today.* Stanford, CA: Stanford University Press.

Glaser, B. & Strauss, A. (1967). *The Discovery of Grounded Theory.* New York: Aldine de Gruyter.

Grunow, D. (1995). The Research Design in Organization Studies. *Organization Science* 6(1):93-103.

Guinan, P. J., Cooprider, J., and Faraj, S. (1998). Enabling Software Development Team Performance During Requirements Gathering: A Behavioral Versus Technical Approach. *Information Systems Research,* 9(2), 101-125.

Guinan, P. Cooprider, J. & Sawyer, S. (1997). The Effective Use of Automated Application Development Tools: A Four-Year Longitudinal Study of CASE. *IBM Systems Journal* 38:124-141.

Hackman, R. (1977). Work Design, in *Improving Life at Work.* R. Hackman & J. Suttle (Eds.) Santa Monica, CA: Goodyear Publishing Company.

Hackman, R. & Oldham, J. (1980). *Work Redesign.* Reading, MA: Addison-Wesley.

Henderson, J. & Venkatraman, N. (1991). Strategic Alignment: A Model for Organizational Transformation via Information Technology. In T. Kockan & M. Useem (Eds.), *Transforming Organizations.* New York: Oxford Press.

Holsapple, C. & Lou, W. (1995). Organizational Computing Frameworks: Progress and Needs. *The Information Society, 11*(1):59-74.

Homans, G. (1967). *The Nature of Social Science,* New York: Harbinger Books.

Howe, K. & Eisenhardt, M. (1990). Standards for Qualitative and Quantitative Research: A Prolegomena. *Educational Researcher, 19*(4):2-9.

Hoyle, R. (1995). *Structural Equation Modeling: Concepts, Issues Applications.* San Francisco: Sage.

Jackson, B. (1987). *Field Work.* Urbana, IL: University of Illinois Press.

Jick, T. (1979). Mixing Qualitative and Quantitative Methods: Triangulation in Action. *Administrative Science Quarterly, 24:* 602-611.

Jones, M. (1999). Mission Impossible? Pluralism and 'Multi-Paradigm' IS Research, in Brooks, L. & Kimble, C. (Eds.) *Information Systems: The Next Generation,* Maidenhead: McGraw Hill: 71-82.

Kaplan, B. & Duchon, D. (1988). Combining Qualitative & Quantitative Methods in Information Systems Research: A Case Study. *MIS Quarterly, 12*(4):571-586.

Kelley, M. (1990). New Process Technology, Job Design, and Work Organization: A Contingency Model. *American Sociological Review, 55*:209-223.

Klein, K. Danserou, F. & Hall, R. (1994). Level Issues in Theory Development, Data Collection, and Analysis. *Academy of Management Review, 19*(2):195-229.

Kling, R. (1987). Defining Boundaries of Computing Across Complex Organizations, in *Critical Issues in Information Systems*, R. Boland & R. Hirschheim (Eds.). John-Wiley & Sons: New York.

Kling, R. (1992). Behind the Terminal: The Critical Role of Computing Infrastructure in Effective Information Systems' Development and Use, in *Challenges and Strategies for Research in Systems Development*, W. Cotterman & J. Senn (Eds.), John Wiley & Sons: London.

Kling, R. & Iacono, S. (1984). The Control of Information Systems Development after Implementation. *Communications of the ACM, 27*(12):1218-1226.

Kling, R. & Scacchi, W. (1982). The Web of Computing: Computing Technology as Social Organization. *Advances in Computers* 21, Academic Press: New York.

Koppel, R. (1994). The Computer System and the Hospital: Organizational Power and the Control of Information, in *Software By Design: Shaping Technology and the Workplace*, H. Salzman & S. Rosenthal (Eds.) New York: Oxford University Press: 143-170.

Langley, A. (1999). Strategies for Theorizing from Process Data, *The Academy of Management Review, 24*(4): 691-710.

Leonard-Barton, D. (1988). Implementation as a Mutual Adaptation of Technology and Organization. *Research Policy, 17*: 251-267.

Lincoln, Y. (1995). Emerging Criteria for Quality in Qualitative and Evaluative Research. *Qualitative Inquiry, 1*(3): 275-289.

March, J. & Simon, H. (1958). *Organizations.* New York, NY: Wiley.

Markus, M. & Robey, D. (1988). Information Technology and Organizational Change: Causal Structure in Theory and Research. *Management Science, 34*(5):583-598.

McClure, C. and Lopata, C.(1996).Assessing the academic networked environment: Strategies and options,Association of Research Libraries, Washington DC.

Merton, R. (1967). *On Theoretical Sociology.* New York: The Free Press.

Meyer, T., Traudt, P. & Anderson, J. (1981). Nontraditional Mass

Communication Research Methods: An Overview of Observational Case Studies of Media Use in Natural Settings, in *Communication Yearbook IV*. D. Nimno (Ed.) Los Angeles: Transaction Books:261-275.

Miles, M. (1990). New Methods for Qualitative Data Collection and Analysis: Vignettes and Pre-Structured Cases. *Qualitative Studies in Education, 3*(1):37-51.

Miles, M. (1979). Qualitative Data as an Attractive Nuisance: The Problem of Analysis. *Administrative Science Quarterly*, 24:590-610.

Miles, M. & Huberman, M. (1994). *Qualitative Data Analysis*. 2nd, Thousand Oaks CA: Sage.

Miller, D. (1991). *Handbook of Research Design and Social Measurement*. Newbury Park CA: Sage.

Miller, D. & Friesen, P. (1980). Archetypes of Organizational Change. *Administrative Science Quarterly*, 25:268-299.

Mintzberg, H. (1979a). *The Structuring of Organizations*. Englewood Cliffs, NJ: Prentice-Hall.

Mintzberg, H. (1979b). An Emerging Strategy of 'Direct' Research. *Administrative Science Quarterly* 24:582-589.

Mohr, L. (1982). *Explaining Organizational Behavior*. San Francisco: Jossey-Bass.

Nadler, G. (1963). *Work Design*. Homewood, IL: Richard D. Irwin.

Orlikowski, W. (1992). The Duality of Technology: Rethinking the Concept of Technology in Organizations. *Organization Science, 3*(3):398-427.

Orlikowski, W. & Baroudi, J. (1991). Studying Information Technology in Organizations: Research Approaches and Assumptions. *Information Systems Research, 2*(1):1-28.

Orlikowski, W. & Robey, D. (1991). Information Technology and the Structuring of Organizations. *Information Systems Research* ,2(2):143-169.

Pacanowsky, M. (1988). Communication in the Empowering Organization. in *Communication Yearbook* 11, J. Anderson (Ed.):356-379.

Pedhauzer, E. & Schmelkin, L. (1991). *Measurement, Design and Analysis*. Hillsdale, NJ: Lawrence Erlbaum Associates.

Pentland, B. (1992). Organizing Moves in Software Support Hot Lines. *Administrative Science Quarterly*, 31:527-548.

Popper, N. (1968). *The Logic of Scientific Discovery*. New York:

Harper Torchbooks.

Robey, D. (1995). Theories that Explain Contradiction. Proceedings of the International Conference on Information Systems, Amsterdam, The Netherlands, SIM Press.

Romanelli, E. & Tushman M. (1994). Organizational Transformation as Punctuated Equilibrium: An Empirical Test. *Academy of Management Journal* 37(5):1141-1166.

Sawyer, S. (Forthcoming). A Market Based Perspective on Software Development. *Communications of the ACM.*

Sawyer, S. (2000a). Packaged Software: Implications of the Differences from Custom Approaches to Software Development, *European Journal of Information Systems, 9*(1), 47-58.

Sawyer, S. (2000b). Studying Organizational Computing Infrastructures: Multi-Method Approaches, in Baskerville, R., Stage, J and DeGross, J. (Eds.) *The Social and Organizational Perspective on Research and Practice in Information Technology*, London: Chapman-Hall, 213-231.

Sawyer, S. (1995). High-Performing Teams and Support Technology in Information Systems Development, Unpublished Dissertation, Boston University School of Management.

Sawyer, S., Cooprider, J. and Guinan, P. (2000). The Social Processes and Performance Patterns of Information Systems Development Teams: A Longitudinal Perspective, Working Paper, School of Information Sciences and Technology, The Pennsylvania State University.

Sawyer, S., Crowston, K., Wigand, R. and Allbritton, M.(2000). Socially Thin and Socially Rich Accounts of ICT use: Evidence from the Residential Real Estate Industry, Working Paper, School of Information Studies, Syracuse New York.

Sawyer S., Farber J. and Spillers R. (1997). Supporting the social processes of software development teams. *Information Technology & People, 10*(1): 46-62.

Sawyer, S. and Gibbons, W. (2000). Implementing Distributed Computing: Parking Gets a New System, in Khrosowspour, M. (ed.) *Annals of Information Technology Cases, 2*, Hershey, PA: Idea Group Publishing: 24-39.

Sawyer S and Guinan P. (1998). Software development: Processes and performance. *IBM Systems Journal, 37*(4): 552-569.

Sawyer, S. & Southwick, R. (1996). Implementing Client-Server:

Issues from the Field, in *The International Office of the Future*, B. Glasson, D. Vogel, P. Bots and J. Nunamaker (eds), New York: Chapman-Hall: 287-298.

Sawyer, S. & Southwick, R. (1997). Transitioning to Client/Server: Using a Temporal Framework to Study Organizational Change. in *Information Systems and Qualitative Research*, A. Lee, J. Liebenau & J. De Gross (Eds.) New York: Chapman-Hall: 343-361.

Schultze, U. (2000). A Confessional Account of an Ethnography About Knowledge Work, *MIS Quarterly, 24*(1).

Shaw, T. & Jarvenpaa, S. (1997). Process Models in Information Systems. in *Information Systems and Qualitative Research*, A. Lee, J. Liebenau & J. De Gross (Eds.) New York: Chapman-Hall: 70-100.

Sica, A. (Ed.). (1998). *What is Social Theory: The Philosophical Debates*. Malden, MA: Blackwell.

Siedler, S. (1973). The Integration of Fieldwork and Survey Methods. *American Journal of Sociology,* 78(6):1335-1359.

Sproull, L., & Goodman, P. (1989). Technology and Organizations: Integration and Opportunities, in P. Goodman & L. Sproull (Eds.), *Technology and Organizations*. New York: Jossey-Bass:254-266.

Sproull, L. & Kiesler, S. (1991). *Connections: New Ways of Working in the Networked Organization.* Cambridge MA: MIT Press.

Star, S. & Ruhleder, K. (1996). Steps Toward and Ecology of Infrastructure: Design and Access for Large Information Spaces. *Information Systems Research,* 7(1):111-134.

Sutton, R. & Staw, B. (1995). What Theory is Not. *Administrative Science Quarterly*, 40:371-384.

Trauth, E. & O'Conner, B. (1991). A Study of the Interaction Between Information, Technology and Society: An Illustration of Combined Qualitative Research Methods, in *Information Systems Research: Contemporary Approaches & Emergent Traditions*, E. Nissen, H. Klein& R. Hirschheim, (Eds.) Amsterdam: North-Holland: 131-144.

Tushman, M. & Romanelli, E. (1986). Convergence and Upheaval: Managing the Unsteady Pace of Organizational Evolution. *California Management Review,* 29(1):29-44.

Van Maanen, J. (1995a). Crossroads: Style as Theory. *Organization Science,* 6(1):132-143.

Van Maanen, J. (1995b). Fear and Loathing in Organizational Studies. *Organization Science,* 6(6):687-692.

Vaughan, D. (1992). Theory Elaboration: the Heuristics of Case Analysis. in *What is A Case: Exploring the Foundations of Social Inquiry*. Ragin, C. & Becker, H. (Eds.). Cambridge, MA: Cambridge University Press.

Weick, K, (1995). What Theory is Not: Theorizing Is. *Administrative Science Quarterly,* 40:385-390.

Weick, K.(1999). Theory Construction as Disciplined Reflexivity: Trade-offs in the 90s, *The Academy of Management Review, (24)*4: 797-807.

Weizenbaum (1976). *Computer Power and Human Reason*. New York: W.H. Freeman and Company.

Wigand, R. Picot, A. & Reichwald, R. (1997). *Information, Organization and Management*. London: Wiley Interscience.

Williams, D. (1986). Naturalistic Evaluation: Potential Conflicts Between Evaluation Standards and Criteria for Conducting Naturalistic Inquiry. *Educational Evaluation and Policy Analysis,* 8(1):87-99.

Wynekoop, J. (1992). Strategies for Implementation Research: Combining Research Methods. in Proceedings of the International Conference on Information Systems, Baltimore: ACM Press, 1992:185-194.

Yates, J. (1988). Control Through Communication: The Rise of System in American Management. Baltimore: Johns Hopkins Press.

Yates, J. & Van Maanen, J. (1996). Editorial Notes for the Special Issue. *Information Systems Research,* 7(1):1-4.

Yin, R. 1989. *Case Study Research*. Beverly Hills, CA: Sage Publications.

Zuboff, S. (1988). *In the Age of the Smart Machine*. New York: Basic Books.

ENDNOTES

1. See OECD, (1998), *The Software Sector: A Statistical Profile for Selected OECD Countries*, Committee for Information, Computer and Communications Policy, Directorate for Science, Technology and Industry, Organization for Economic Co-operation and Development, Report DSTI/ICCP/ AH(97)4/REV1 at http://www.oecd.org/dsti/sti/it/infosoc/stats/software.htm (Sep. 20, 1999).

2. I invoke these scholars' names to point out their use of reflection

and to highlight the value which they place on reflection. In essence, I appeal to higher authority! A naive reader might think I am setting my work near theirs and this is not so.

ACKNOWLEDGMENTS

This chapter has benefitted greatly from comments by, and/or discussions with, Eileen Trauth, Angela Barnhill, Zuzana Sasovova, Don Falconer, Kristin Eschenfelder, Matthew Jones, M. Lynne Markus and two anonymous reviewers.

APPENDIX 1: EXPLANATORY MATRICES

In an explanatory event matrix the themes form one axis and the sources of evidence form the other axis (Miles and Huberman, 1994; Miles, 1990). The cells formed by this matrix contain pointers back to the source of evidence relative to the theme (or concept) to which it relates. As an example here is (an abridged) part of an explanatory matrix used in Sawyer (2000b).

Themes\Evidence	Mgr1 Interview1	Mgr2 Interview1	Mgr2 Interview2	Meeting1 Observation	PM Plan N
Specific Implementation Characteristics	*Note: This group of characteristics reflects issues with implementation. Evidence supporting each of the four characteristics is drawn from a range of sources.*				
Multiple stakeholders	See lines 4-10.	See lines 55-57, 61-64, 102-105.	See lines 45, 49, 76-81.	See the diagram and field notes, pp. 2 & 4.	See sections on stakeholders and steering committee.
Multi-level	See lines 53-75.	See lines 83-92, 115-132, 134.	See my comments to this interview.		
Multiple effects	See lines 111-138.	See lines 151-157. See the two memos MGR2 provided with this interview.		See field notes pp 7-9.	See risk management section (and updates).
User schedules				See field notes p. 10	See plan v. updates.

Issues
for the
IS Profession

Chapter VIII

Conducting Action Research: High Risk and High Reward in Theory and Practice

Richard Baskerville
Georgia State University, USA

INTRODUCTION

Action research is a method that solves immediate practical problems while expanding social scientific knowledge. Based on collaboration between researchers and research subjects, it is a cyclical process that builds learning about change into a given social system (Hult & Lennung, 1980). On the surface, the discipline of information systems (IS) would seem to be a very appropriate field for the use of action research methods. IS is a highly applied field, almost vocational in nature (Banville & Landry, 1989). Action research methods are highly clinical and place IS researchers in a "helping-role" within the organizations being studied (cf. Schein, 1987, p.11). It should not be surprising that action research has been characterized as the "touch-stone of most good organizational development practice" and that it "remains the primary methodology for the practice of organizational development" (Van Eynde & Bledsoe, 1990, p. 27). Action research merges research and praxis, thus producing relevant research findings.

Such relevance is an important measure of the significance of IS research (Keen, 1991).

However, the action research method has proved unpopular among IS researchers, particularly North American IS researchers. Action research articles in major North American research publications are disproportionately rare. Orlikowski and Baroudi (1991) discovered only one action research article among the 155 major research publications between January, 1983 and May, 1988. While action research represented only a tiny fraction of major IS research articles in the mid-1980s, a longitudinal study extending across the 1970s, 1980s and 1990s revealed a steadily rising number of significant IS action research articles (Lau, 1997). Of the 30 IS action research journal articles published between 1971 and 1995, nearly half of these were published in the 1991-1995 time period. Although the successful use of the technique is attracting increasing IS journal attention, this is still a tiny fraction of the IS research corpus. Despite its overwhelming acceptance in organizational development, it is a poorly represented approach in IS, particularly among North American IS researchers.

Outside of North America, action research has made more contributions to the literature of the IS research community. Peter Checkland's Soft Systems Methodology (Checkland, 1981) and Enid Mumford's ETHICS (Mumford, 1983) both evolved from and incorporate action research. Their work has heavily influenced IS research by linking action research and systems development. This linkage has increased the presence of action research in the British, Scandinavian and Australian IS literature. However, action research is not a predominant IS research method even in those geographic regions.

Given the potential in the relationship between the vocational nature of the IS field and the clinical nature of action research, why is action research contributing so little to the IS research literature? One serious aspect of the explanation lies in the substantial risks in conducting action research. The risks associated with action research are not typically the explicit focus of published discussions of the techniques.

Consequently, the goal of this chapter is to define and analyze these risks and to discuss their practical effects. An understanding and awareness of the risks enables the action researcher to manage these risks, particularly in terms of an academic career. In order to keep this analysis within a useful sphere, the discussion will be grounded in the context of a range of six actual action research cases.

AMBIGUITY AND RISK

For a researcher, ambiguity is risk. Ambiguity refers to a concept that is undecided, an idea that is open to interpretation in more than one way. Researchers who allow too much ambiguity in their purpose, their methodology and their learning may be unable to produce knowledge with enough definition to be useful. Ambiguity is a problem for the action researcher from the moment the approach is adopted. Usage of the very term "action research" is loaded with ambiguity. The term "action research" is used, on the one hand, to refer both to a general class of methods in social enquiry, and on the other hand, to a specific subclass of those methods as distinguished from "action science", "action learning", "participatory action research", etc. This terminological ambiguity arises because action research began as a unified approach to social enquiry and then fragmented throughout its history (Baskerville, 1999). In its origins, the essence of action research is a simple two-stage process:

> First, the diagnostic stage involves a collaborative analysis of the social situation by the researcher and the subjects of the research. Theories are formulated concerning the nature of the research domain. Second, the therapeutic stage involves collaborative change experiments. In this stage changes are introduced and the effects are studied (Blum, 1955).

A more precise definition of IS action research is drawn from the published characteristics of action research in the social science literature (Baskerville & Wood-Harper, 1998). However, this literature is dominated by the particular five-process form of action research described by Susman and Evered (1978), and tends to emphasize action research characteristics based on goals and objectives rather than characteristics based on the process. Adapting Hult and Lennung's definition (1980), four major characteristics of IS action research are distinguishable[1]:

1. Action research aims at an increased understanding of an immediate social situation, with emphasis on the complex and multivariate nature of this social setting in the IS domain.
2. Action research simultaneously assists in practical problem solving and expands scientific knowledge. This goal extends into two important process characteristics: first, there are assumptions being made that observation inevitably subsumes interpretive

acts; second, the researcher intervenes in the problem setting.

3. Action research is performed collaboratively and enhances the competencies of the respective actors. A process of participatory observation is implied by this goal. Enhanced competencies (often the result of collaboration between diverse professionals) is relative to the previous competencies of the researchers and subjects, and the degree to which this is a goal, and its balance between the actors, will depend upon the setting.

4. Action research is primarily applicable for the understanding of change processes in social systems.

Action research is sometimes confused with participatory observation itself. Participatory observation is a data-collection technique in which the research observations are taken by one of the research subjects. While necessarily a data-collection technique in action research, this technique is useful in case studies and ethnographies that may lack the other defining characteristics of action research, e.g., problem solving, change process focus, competency enhancement, etc.

PHILOSOPHICAL AND METHODOLOGICAL RISKS OF ACTION RESEARCH

Action research is not only an obscure term, but the basic ideas are found in philosophical positions that are deemed to be questionable by many natural and positivist scientists. Action researchers risk rejection of their findings by some social scientists because of the conflict with traditional scientific values such as deductive logic, problem reduction, and detached observational objectivity. Action researchers generally accept that complex social systems cannot be reduced for meaningful study. Human organizations, as a context that interacts with information technologies, can only be understood as whole entities. Factoring of a social setting, like an organization and its information technology, into variables or components, does not lead to relevant knowledge about the real organization. For the action researcher, complex social processes can be studied best by introducing changes into these processes and observing the effects of these changes.

This change-orientation is the foundation of action research. Kurt Lewin developed the original action research approach as a form of clinical, action-based social psychology (Lewin, 1947a, 1947b). It is based on the fundamental pragmatist idea that a concept can only be clearly understood by its consequences. Theories are only validated by their practical results. Right down to these foundations, action research involves some risky assumptions. Pragmatism itself has never been an entirely respectable form of thought because some feel it leads to unprincipled expediency (Thayer, 1981).

Once the researcher allows social intervention into the research setting, then an interpretivist perspective on data must replace the positivist perspective of detached, objective observation. When researchers intervene, they become part of their own group of study subjects. Conducting action research empirics involves interpretive statements that explicitly permit intrusion by the observer's values and *a priori* knowledge. Researchers seek the "meaning" in their observations, and accept the Kantian notion that individual understanding colors observation data and consequent deductions. Further, social interaction with the subjects will yield reflection that will add another layer of color to data.

Following from the interpretive data, action researchers face risks that other scientists will challenge their underlying data and analytical techniques. Acton researchers most often use qualitative data because of their interpretive assumptions. This "soft" data cannot be legitimately analyzed in its original state with the full set of quantitative operations without further qualitative interpretation through mapping, indexing and scaling (Halfpenny, 1979). Qualitative analysis adopts techniques like hermeneutics and deconstruction, which to some social scientists seems more appropriate for historical and literary analysis than social science (cf. Baskerville & Pries-Heje, 1999).

Lastly, because the conduct of action research will typically seek an in depth study of the experiences of a specific organization, researchers run afoul of the statistical notions tightly held by positivist social scientists. The action researcher risks criticism on the basis of traditional analytical techniques like representative samples and statistical generalization, along with traditional notions of validity like repeatability, reliability and falsifiability.

It can be seen that those conducting action research adopt a risky position because their philosophy and their method of inquiry are subject to challenge from the alternative perspectives of traditional social science philosophy. The pragmatism, holism, interpretivism, idiography, qualitative data and qualitative analyses are basic sources of unease to traditional social scientists. There is consequently an essential risk inhabiting the very assumption space of the technique. This risk is that the findings will be rejected and ignored by large, important segments of the social science community.

EXAMPLES OF ACTION RESEARCH

A good way to consider practical risks in conducting action research is to study the inherent risk profiles of several action research examples. For illustration purposes, six action research projects are selected from those invited for discussion at the 1998 North American Information Systems Action Research Workshop in Atlanta, Georgia, USA. This section provides a brief overview of these six action research projects describing the clients, the nature of the interventions and the findings of the studies. After this section, the examples will provide a context for discussing the nature of the practical risks. So this section will provide an introduction and overview, while further relevant details of the examples will be examined throughout the remainder of this chapter.

"Semantic Database Prototypes" (Baskerville, 1993)

This research was conducted through a consortium of several American universities and involved a government information systems development project that had failed. The study team faced the development of a procurement budgeting system for a government agency after the collapse of two previous analysis and design efforts. The team was composed of an experienced team leader with a strong practical background in logistics and information systems, an analyst with a strong background in the procurement system, a scientist commissioned as an action researcher, and later included a programmer and a second analyst. The study applied a number of IS theories, with varying success, to the problem. Finally, they developed a theory of semantic consensus, a condition where there is pragmatic agreement among database designers and all of the users about which aspect of reality is being represented by a particular database element, leading to a successful design experience.

"Revealing Complexity in ISD" (Chiasson & Dexter, 2001)

In this project, action research was used to examine the development of an electronic medical record system in two heart clinics over a four-year period. The primary researcher was involved directly in starting, developing and implementing the software and infrastructure in the two heart clinics. The research discovered a number of epochs of specific technology-context processes extracted from daily events. These epochs arise when sudden, complex processes (like shocks and conflict) unexpectedly altered the project direction. Sources of these complex processes included technology-supplier influence, use of prototyping, and longer time horizons. Using this source-complexity-epoch theoretical framework, a satisfactory implementation of the patient record system followed.

"An Action Research Study of Asynchronous Groupware Support" *(Kock & McQueen, 1998)*

This study arose when the researcher was asked to intervene following an unsuccessful attempt to adopt a conferencing application to support strategic planning. The researcher joined seven process redesign groups as both a facilitator and a researcher into the issue of the effects of asynchronous groupware on outcome quality and productivity in process redesign. One process redesign group was studied over a period of one month and six process redesign groups were studied over a period of four months. Data was collected through participant observation, interviews, and automatic computer generation of transcripts of electronic group discussions. The data included field notes, tapes and transcriptions. The study found that asynchronous groupware could increase productivity considerably. There was a balance between negative and positive factors that limited increases in quality. The theory describes a system of positive effects, positive consequences, negative effects and negative consequences. Based on this theory, a development methodology was created and applied. The practical result was a complete groupware-supported strategic planning project.

"Building a Virtual Network" *(Lau & Hayward, 2000)*

This research incorporated a two-year project to build a virtual network as part of a research-training program in community health. The study used action research to build the virtual network as an integral part of the research-training program. The researchers intervened by suggesting the technology to be adopted, facilitating the use of certain software tools and information resources, collecting research data for analysis, and offering feedback to the participants and staff on a periodic basis. The underlying theory was derived from the principles of virtual teams with an emphasis on the role of technology in establishing this network. The program enrolled 25 participants from 17 health regions. These participants had responsibilities to design and conduct research projects, provide information to support decision making, and act as an expert resource on community health issues. The collaborative research team included nurses, planners, and community health research officers with the practical result of a successful virtual network.

"Exploring IT Support for Organizational Learning in the Virtual Corporation" (Nosek, 1998)

The study process in this research involved researcher-practitioners (called "evaluators" in the study) comprised chiefly of executive MBA students. These evaluators were mostly mid-level to senior-level managers and many actively participated in these decisions. These evaluators were introduced to a number of collaborative technologies. Using a theoretical foundation of group sensemaking and organizational interpretation, they analyzed organizational decisions with respect to how information technology supported (or could have supported) critical decisions. These evaluations were done individually by evaluators due to the sensitivity of the information and small groups were formed to identify any general similarities or differences in the processes and the way that information technology was used (or could have been used). Evaluators presented their findings to the other executives involved in the actual decisions for feedback regarding the possible use of information technologies. These executive-action-researcher-evaluators were reflecting on their decision processes and gathering information to make sense within their given situation. Their "insider" status developed a candidness that was unprecedented. The report was able to maintain a critical level of anonymity focused on exemplars of decision formulation and direction setting, and how information technology could support decision processes. The practical outcomes were the improved decision processes of these decision-makers.

"Coping with Systems Risk" (Straub & Welke, 1998)

This research included an intervention in one large firm with information technology services. The theory proposed in the study was that proper education of professionals could transform their thinking about security policies in information systems. Prior to the intervention, single-tiered, preventative security options dominated the thinking. After the intervention, policies that had a multi-tiered and deterrent impact were actively pursued. More than 25 top managers, middle managers and other professionals were involved in the comparative study, which took place over a 15-month period. The involvement included a security planning team charged to investigate the firm's information security options. The action research introduced theory-based planning models for strengthening security with the result that managers considered a wider range of security safeguards. The practical outcome was a measurable improvement in the range of security options being considered by the analysts.

PRACTICAL RISKS WITH ACTION RESEARCH

At a more functional level, risk involves exposure to the chance of injury or loss. Beyond issues of ambiguity, philosophy and scientific methodology, do action research projects constitute a "dangerous chance" being taken by researchers or the client organizations? Primarily, such a dangerous chance implies that there may be an "outcomes failure" from undertaking an action research project. Outcomes failures result when action research fails to achieve its stated purposes or quality in its processes. Conversely, I characterize the successful conduct of action research project as one that does not suffer outcomes failure. Assuming action research risks are similar to other kinds of risks, a research methodology risk study can be built in a manner similar to systems development risk studies. Using systems development risk as a reference, I discover four elements in these risks: (1) the types of outcomes failure, (2) the degree of probability of an outcomes failure, (3) the amount of damage or impact of the outcomes failure, (4) the stakeholders affected by the outcomes failure (Baskerville, 1991). I will organize the discussion of these four elements into three sections. First I will examine potential outcomes failures using a review of published action research philosophy. Second, I will consider the degree of probability and the amount of damage in a section on risk levels based on the real-world examples of information systems action research described above. Third, I will examine the effects on stakeholders illustrated by the examples.

Outcomes Failure

The practical risks of conducting action research projects may be ultimately summed into the possibility that the central goals of action research will fail to achieve a significant result. In general, any research methodology can engender an outcomes failure. However, for action research in particular, I characterize outcomes failure in two degrees.

Second-Degree Outcomes Failure

A *second-degree outcomes failure* arises when the action research project achieves a solution to the immediate problem setting without any significant contribution to scientific theory. Second degree action research outcomes failures may be thought of as "failures of form." For example, the ultimate theoretical basis for solving the problem may be well established and widely known. While the project adds further confirmation to this theory, it would fail to achieve any significant scientific contribution[2]. This failure is a catastrophe for a scholarly researcher because there is little chance for publishing significant scientific results.

For example, in Nosek's "Exploring IT Support for Organizational Learning in the Virtual Corporation" (1998), the theoretical constructs underlying the action were based chiefly on published theory. While the study (as written) confirms existing theory, and makes a contribution to that degree, this study is not as publishable as one that develops original theory. In contrast, in Straub and Welke's "Coping with Systems Risk" (1998), an original theory-based model for strengthening security resulted. This model was synthesized from a wide variety of "intellectual tools" in the security literature and models of managerial decision making. The action research demonstrated the usefulness of this newly developed and original model in action.

First-Degree Outcomes Failure

A *first-degree outcomes failure* arises when the practical problem setting remains unresolved at the conclusion of the project. A first-degree outcomes failure is more catastrophic because theoretical validity of action research results is anchored in the resolution of a practical problem. If the project concludes without solving the practical problem, it becomes necessarily difficult to validate any theoretical contribution proceeding from the project. Hence, neither practice nor science is served.

There are no good examples of first degree outcomes failure among the work highlighted above. However, we can see subtle differences in the ultimate practical impact of these outcomes. For example, in my own article (Baskerville, 1993), the practical outcome was a *design*, not a working system.[3] Similarly, in Straub and Welke (1998), the practical outcome was improved security *planning*, while the actual improvement to security itself is not shown. On the other hand, Lau and Hayward's study (2000) and Chiasson and Dexter's (2001) work resulted in creation of a virtual training network and implementation of a working client record system, respectively. The practical outcomes depend on how those conducting action research define the practical problem being targeted. Our study team defined an analysis and design problem and Straub and Welke's defined a management problem. Lau and Hayward, and Chiasson and Dexter, defined systems implementation problems.

No doubt action research teams learn from first degree outcomes failure, and it seems intuitively appealing that this knowledge "learned from our failure" ought to be equally valuable to knowledge "learned from our success." In practice, a failure to improve a problem setting, especially a worsening problem setting, amounts to a non-effect. The pragmatic roots of action research respect only *consequences* as the basis for clear understanding of concepts. The lack of consequences results from cloudy thinking and poor understanding. Owing to this essential philosophical precept, conducting action research is particularly risky because first-degree outcomes failures are research failures as well as practical failures.

Risks to Stakeholders in Action Research

There are two main groups of stakeholders traditionally distinguished in action research projects: clients and researchers. Clients are most concerned that the immediate practical problem is resolved. A second-degree outcome failure is only of indirect concern to the clients. Researchers are also concerned with resolution of the underlying practical problem, however, their interests are most directly served by a significant contribution to scientific theory. Hence, a second-degree failure concerns mainly researchers. A first-degree outcome failure is the researchers' nightmare. The client is necessarily disappointed in the researcher. But also, a first-degree outcome failure means a wasted commitment of research time and resources in a project that made no significant contribution to science and solved no practical problems.

The risks to the client and researcher are usually interlocked, at least to the extent that a first-degree outcomes failure is undesirable. However, the additional interest of the researcher in avoiding a second-degree outcomes failure may distinguish different risk profiles for client and researcher (See Table 1). Two studies in this selection illustrate differing profiles between client and researcher stakeholders. In "Revealing Complexity in ISD" (Chiasson & Dexter, 2001), a creeping commitment by the client and growing hostility in the client relationship inevitably joined the originally separate risk profiles. Originally, the client organization was only minimally interested in the research outcomes, essentially patronizing the researchers at the outset. The researchers, however, made substantial commitments to the project, including dependence of a degree thesis on the avoidance of second-degree outcomes failure. As the project evolved, the client became

Table 1: Risk Characteristics of the Examples

Example Risk	Client Risk	Researcher
"Semantic Database Prototypes" (Baskerville, 1993)	Substantial	Substantial
"Coping with Systems Risk" (Straub & Welke, 1998)	Substantial	Moderate
"An Action Research Study of Asynchronous Groupware Support" (Kock & McQueen, 1998)	Substantial	Substantial
"Building a Virtual Network" (Lau & Hayward, 2000)	Minimal	Minimal
"Revealing Complexity in ISD" (Chiasson & Dexter, 2001)	Rising	Substantial
"Exploring IT Support for Organizational Learning in the Virtual Corporation" (Nosek, 1998)	Minimal	Minimal

increasingly interested in the results, and increasingly dependent on a successful problem resolution outcome. During this period, however, conflicts made the conduct of action research even more risky. Thus the client exhibited a rising risk profile while the researchers maintained a substantial risk profile more or less throughout the project.

In "Coping with Systems Risk" (Straub & Welke, 1998), the critical nature of information security problems in the client organization made this project a substantial risk for the client organization. The potential amount of damage and impact of problems resulting from the study were clearly very important to the client. The researcher, however, made a more reserved commitment to the study. This reservation does not imply that the researcher was less committed to avoiding first degree outcomes failure. However, the research design was constructed in such a way that a practical success might be possible, even if the results of the action yielded a first-degree outcomes failure. This success was made possible by the case study component in the research, which opened the possibility of a valid research outcome from the case element of the study, even if the action element failed. Hence, the client organization exhibits a "substantial" risk profile, while the researcher exhibits a "moderate" risk profile.

Underlying Risks Leading to Outcomes Failure

But outcomes failure alone is not the only element of risk in conducting action research. Looking deeper, there are inherent causes of action research outcomes failure, and these underlying risks leading to outcomes failure should be further examined. The most obvious aspects of risk traditionally regarded in the action research literature are the dilemmas confronting the researchers. Rapoport (1970) originally described these dilemmas in terms of "ethical", "goal" and "initiative" dilemmas. Warmington (1980) later characterized these as "value", "goal" and "role" dilemmas. Closer examination of these dilemmas reveals at least four types of underlying risks in conducting action research.

Completion Risk

There is a risk that researchers will disengage from the research (i.e., quit) before they achieve a core understanding of their actions. This early disengagement has implications for the client organization because the early disengagement may mean that the effects of the practical actions will prove short-lived. There are also implications for the researcher whose theoretical framework may contain errors because it is founded on an incomplete understanding of reality in terms of the ultimate consequences of their action. The situation can develop where the researcher declares "success" and leaves at the first concrete indication of improvement, instead of waiting to see if the improved situation will endure. Thus there is a risk to the researcher that an uninformed theory may propagate. The uninformed theory may not fully incorporate the substantive ends toward which the action was taken. It may neglect some aspect of the universe of moral concerns that inspired the action. This problem is seen in Susman and Evered's (1978) substitution of "understanding" as a preferable criterion to "explanation". It also relates to Rapoport's (1970) ethical dilemma in the sense that the researcher may disregard the ethics of the subjects, wrongly assuming that an ethical framework is shared by researcher and subject. As a result, there may be disagreement over whether the core problem has been resolved.

Domination Risk

There is a risk that the conducting of action research may proceed without true collaboration between subjects and researchers. Classic action research theory accepts that the researchers must be coauthors of the stimuli that make things happen in the research. They can neither be the sole authors, nor be excluded from intervening in the subject organization. They must collaborate as co-producers of the intervention because of the goal dilemmas (Rapoport, 1970; Warmington, 1980) that confront either the research subjects or the researchers in isolation from each other. In other words, the researchers' dominant goal is necessarily epistemological; the subjects' dominant goal is directly practical. The researcher wants to create general knowledge; the subject wants to fix their practical problem. If the researchers' motives dominate the process of inciting action, there is a risk that the actions may not be guided closely enough by the immediate problem situation. If the subjects' motives dominate this process, there is a risk that there will be a weak theoretical basis for the actions. If either party dominates the process of inciting actions, ultimately the risk increases that the understanding of the effects of the action is defective. Basic pragmatist philosophy rejects any theoretical concept that does not rest on clear understanding of the effects of the concept's use.

Detachment Risk

Because those conducting action research are guided by immediate practical problems, their close engagement with their subjects is essential. As I mentioned earlier, action research substitutes engagement for the objective detachment of positivist science. The conduct of action research presupposes participant observation, which exchanges the subjectivity risks of positivist science for the role dilemmas of action research. Primarily this role dilemma regards the great difference between the professional obligation to the client in solving the immediate problem, and the scientific obligation to explore imaginative new theories (Trist, 1976). This role dilemma further presupposes an extremely delicate balance between collaboration and direction, experimentation and expertise, professional and scientific priorities, and client-system or research loyalties (Clark, 1976). If researchers become detached from either their scientific or practical goals, there is a risk that the research will fail to achieve any significant advances either scientifically or practically.

Abstraction Risk

There is also a delicate balance between action and reflection in action research. When the problem setting is complex and risky, researchers may be further drawn into extensive study and reflection in order to minimize the potential harm of poorly considered action. Extensive abstract study periods are not consistent with action research principles. As a substitute, those conducting action research iterate theorizing with action and outcomes evaluation. In a manner similar to system prototyping, action research involves testing an imperfect "prototype" theory by stimulating action in accordance with the theory. If the results are unsatisfying, the theory is changed. Prolonging any periods of abstract contemplation destroys the relevance of the theory by increasing the gap between theory space and reality space. This risk is related to Warmington's (1980) value dilemmas because researchers are trained to value the contemplative act, while practitioners value swift, decisive action. This risk is also related to Rapoport's (1970) initiative dilemma because the allure of scientific research ideals draws researchers to initiate research by reflectively choosing their subjects. The regard for abstraction is different in objective methods where the researchers retain control over the basic topic and can reflectively choose what range of phenomena to observe in their subjects. In conducting action research, the topic must sometimes swiftly follow the practical setting. This risk is also similar to risk avoidance behavior in the practical work that can sometimes freeze decision-makers.

These four types of risk—completion, domination, detachment and abstraction—may contribute to producing either of the two possible degrees of outcomes failure.

RISK LEVELS IN ACTION RESEARCH

How high are the risk levels for those conducting IS action research? I will examine the risk levels of the various stakeholders in the action research examples described above. Differing levels of risk between the two main stakeholder groups (researchers and clients) characterize these risk levels. Each of these characteristics is discussed below, illustrated by the six action research project examples. Table 1 summarizes the characterization of these examples and is discussed in more detail below.

Minimal

A minimal risk level suggests that there is a low degree of probability that problems will develop with completion, domination, detachment, or abstraction to the degree that a serious outcomes failure will develop. Minimal risk accordingly implies a high degree of probability that the project will reach a significant theoretical and practical contribution. Because action research projects are sometimes initiated in settings of substantial uncertainty, this risk level may be unusual in action research projects. Nevertheless, I see two examples that appear to involve minimal risk strategies. In "Exploring IT Support for Organizational Learning in the Virtual Corporation" (Nosek, 1998), an outcomes failure was less threatening because the action research project was not entirely visible to the client organizations, nor crucial to a formal research organization. The more-or-less informal arrangement with the executive students allowed the action researchers to carry out the research in conjunction with their normal duties. While this arrangement implied a higher degree of completion risk, in that the subjects could easily choose to break off pursuit of the action research goals, the probability of domination, detachment and abstraction risk levels were low. In addition, the infrastructure involved a minimal investment by the researchers in terms of extra time and resources. The perception of damage or impact as a result of outcomes failure would be minimized by the low visibility of the project.

In "Building a Virtual Network" (Lau & Hayward, 2000), the setting was more formal and involved much higher visibility both to the client and research organizations. However, this project also exhibits minimal risk characteristics because the context is constructed in a cautious and conditional manner. This context is presented to the client as a "pilot project" intended to determine feasibility, and the action research component seems a relatively minor appendage to the project. As a consequence, a first-degree outcomes failure is unlikely, since overall feasibility is likely to be indicated (i.e., it will be shown either feasible or infeasible) as a result of the project success or failure. However, there is a higher probability of second-degree outcomes failure than that in the "sensemaking" study, and chances of completion, domination, detachment and abstraction problems are higher. Still, the potential for damage or negative impact is low, and this low potential attenuates the impact of the failure of the research elements of the pilots. These elements involved the relatively incremental effects of adding group technologies into the pilot project. The impact or damage from a second-degree outcomes failure is minimized by this fairly mild resource commitment by the researcher to the action research aspect of the project.

Moderate

A moderate risk level suggests that there is a more substantial degree of probability that problems will develop with completion, domination, detachment, or abstraction. Moderate risk accordingly implies a lessened degree of probability that the project will reach a significant theoretical and practical contribution. One example in Section Two appears to involve moderate risk strategies. In "Coping with Systems Risk" (Straub & Welke, 1998), the action opportunity appears to arise more-or-less unplanned during the course of a case study project, affording the opportunity to intervene in the case and introduce a training program for top managers. The original case study was a high visibility project within the client organization, and seizing this opportunity has introduced new probabilities for outcomes failures. The action research component involved a heightened degree of uncertainty because of its unplanned genesis. Completion, domination, detachment and abstraction risks could be characterized as "typically high" for action research. What distinguishes this project as moderate risk is the emerged design, in which the action component was important, but not completely essential to the research. However, for the client, the action component was critical for the practical outcomes. The result of the case study would have provided only a very limited degree of practical improvements without the action research component. Mostly, the case study would have provided feedback to the organization regarding its problem setting. The action research is an incremental advance within the research strategy. While the probability of risk is at the normal high levels for action research, the amount of damage or impact was limited.[4]

This research design aspect has little effect from the perspective of the client organization. The triangulated research design (conducting a case study, and then conducting action research) would certainly improve the possibility of a practical solution. From the client's perspective, their focus was this practical improvement. This focus means that their risk profile was not reduced by the triangulated research methodology. Their substantial commitment of executive and professional time along with their practical problem of information security remained substantially at risk regardless of the research design.

Substantial

A substantial risk level suggests that there is a high degree of probability that problems will develop with completion, domination, detachment, or abstraction to the degree that a serious outcomes failure will develop. Substantial risk accordingly implies a low degree of probability that the project will reach a significant theoretical and practical contribution. Because action research projects are sometimes initiated in settings of substantial uncertainty, this risk level may be usual in action research projects. Two of our examples appear to involve these typical, substantial risk profiles. In "An Action Research Study of Asynchronous Groupware Support" (Kock & McQueen, 1998), the project involves teams of process redesign groups. These groups represent a substantial resource investment on the part of the client organization. From their perspective, this was a project with high visibility and high risk. There was also a considerable commitment of time by the researcher. These are high stakes, and the amount of damage and impact of an outcomes failure would be dramatic. Likewise, the dependence of the project on groupware technology, itself an emerging technology with mixed outcomes, particularly raises the degree of completion risk and domination risk. This project exhibits a substantial risk profile. Similarly, in "Semantic Database Prototypes" (Baskerville, 1993), a formal client-system infrastructure between government agencies, along with previously failed, expensive attempts to resolve the immediate practical problem, raised the visibility of the action research project. This project was also high stakes, with a large commitment of resources from client and research organizations. From the client's perspective, this further commitment of resources in the wake of previous project failure meant a lot was at risk. This risk setting is especially severe because the inter-organizational team increased the probabilities of domination and detachment risk.

Rising

Action research projects are necessarily emergent, and the risk profile of a project is likely to evolve. A rising risk level suggests that there is initially a low degree of probability that problems will develop with completion, domination, detachment, or abstraction to the degree that a serious outcomes failure will results. However, over time, the impact of failure may grow more serious with increasing investment in the project by the client or the researchers. The setting may broaden in scope such as to encompass increased risks that completion, domination, detachment, or abstraction problems will arise. Rising risk implies a falling degree of probability that the project will reach a significant theoretical and practical contribution. Because the outcome will necessarily become more or less apparent as any failing action research project approaches its conclusion, it is difficult to distinguish any failed action research projects from projects with rising risk. However, I retain the category to distinguish projects in which some event clearly changed a project's risk profile from one category to a higher category. One example clearly exhibits a rising risk profile: "Revealing Complexity in ISD" (Chiasson & Dexter, 2001). In this project, two events affected the risk profile. In one, conflicts erupted between clients and researchers. The likelihood that researchers might withdraw partially or completely from the project arose. This conflict increased the probabilities implicit in completion, domination and detachment risks. The other event was the creeping commitment of the client in terms of resources dedicated to the project. As the project absorbed more client time and resources, it became increasingly important to avoid a first-degree failure. Thus, from the client's perspective, the resources at risk were rising, as was the detachment risk because of the conflict. In risk terms, the potential amount of damage or impact of a second-degree outcome failure rose steadily from minimal (casual interest) to substantial (major commitment).

Falling

It might also be assumed that the counter situation could arise with respect to rising risk. That is, emergent action research projects may exhibit an evolving risk profile that entails progressively lower risk levels. A falling risk level suggests that there is initially a high degree of probability that problems will develop with completion, domination, detachment, or abstraction risks to the degree that a serious outcomes failure will develop. However, over time, the problem setting may clarify and the impact of failure may grow less serious with the possibility of quick and inexpensive solutions to the problem setting. Perhaps an interesting new theory may also appear that rapidly addresses the problem setting. The setting may suddenly close in scope such as to eliminate many risks that completion, domination, detachment, or abstraction problems could arise. Falling risk implies a rising degree of probability that the project will reach a significant theoretical and practical contribution. Because the outcome will necessarily become more or less apparent as the conduct of any successful action research project approaches its conclusion, it is difficult to distinguish any successful action research projects from projects with falling risk. However, I retain the category to distinguish projects in which some event clearly changed a project's risk profile from one category to a lower category. There are no examples in my selection of action research cases of these types of risk profiles, but I believe it is likely that there is potential for such projects to occur.

CONCLUSION

I will conclude this chapter with some brief practical advice. First, some advice based on how I believe one should go about managing these action research risks discussed above. This advice is based purely on my experience, and is rendered with the caveat that every action research project is unique—there are no panaceas. Second, I would like to reflect on conducting action research and the inherent risk to an academic career. I believe that the academic culture in IS research has led to a level of risk-avoidance in academic careers that is unhealthy for the field. Higher-risk, but potentially higher-reward approaches such as action research are being avoided far more than can be justified.

Managing Action Research Risk

From the risk profiles above, it appears that it may be possible to effectively manage the risks in conducting an action research project.

Managing the Types of Outcomes Failure

The inventory of risk outcomes and types of risk helps us to qualitatively measure the degree of risk in conducting action research, and to recognize when this degree is changing. Maintaining a distinct focus on the failure potential from both a first-degree and second-degree outcome may help recognize rising risk sooner. It is possible to control completion risk, domination risk, detachment risk and abstraction risk more effectively once their effects are understood and their presence qualitatively appraised.

Managing the Probability of Outcomes Failure

To a large degree, probability of first or second-degree outcomes failure will be a tradeoff in terms of resources committed to the project. Increasing resources will often lower the probability of failure. Unfortunately, this solution is a double-edged sword. This commitment increases the amount of damage and impact of a failure if one does result. An alternative technique is to reduce the scope of the expected practical and/or theoretical outcome. This reduction has three effects: first, it may lessen the resource demands and thereby reduce the damage from a failure outcome. Second, it may reduce the complexity of the practical problems being addressed by the project. Third, the theoretical focus may actually be improved by a reduced practical scope. The reduced problem may permit actions to be more tightly linked to a particular theory, and help limit the range of alternative explanations for the effects of action.

Managing the Amount of Damage and Impact of Outcomes Failure

Early in the research, and at regular intervals, evaluations should be undertaken to determine the effects of first or second-degree outcomes failure. These evaluations should focus on the impact of a failed study on both the client and the researcher. In situations where the probability of failure is uncontrollably high, the stakeholders may create contingency plans and prepare their execution in order to lower the damage. The net effect of such "escape routes" is to lower the risk profile of the action research project itself.

Managing Stakeholder Risks

Maintaining an even level of risk among the stakeholders may be one of the most effective means for managing risks. Action research projects that are conducted in the beginning without the interest and commitment of either the researcher or the client may run a high degree of completion risk. Action research participants should also recognize when the risk profiles of the stakeholders is becoming progressively unbalanced during a project. Thus, if the client is being increasingly placed in more substantial risk than the researcher, the project participants should take steps to reduce the risk profile of the client, increase the risk profile of the researcher, or both.

Academic Career Risk

In my discussion of the risks inherent in conducting action research, and the mechanisms for managing these risks, I have not addressed one of the most serious inhibitors to the conducting of action research. This inhibitor is embodied in the academic research culture of "publish or perish." The process by which action research has failed to gain the mainstream in the IS research literature follows a typical history. In the early stages of an academic career, i.e., those of a doctoral student, most academics are encouraged to avoid risks. Particularly in U.S. programs, there is great pressure on the student to quickly construct an acceptable research thesis. Consequently, riskier techniques such as action research tend to be eschewed in favor of safe, mainstream techniques like experiments or questionnaire surveys. Such techniques are safe because they entail a predictable time involvement and can be constructed in a way that a contribution to knowledge is almost an inevitable result. This safe and expedient route becomes advised and selected. However, once a doctoral candidate graduates and joins the ranks of junior faculty in universities, the pressure actually increases to continue the established line of research from the doctoral work. There is pressure to refine the thesis material into a journal article, and to build upon the individual's established research framework to gain a train of several respectable publications. Essentially, the junior researchers become entrapped within their own expediencies as doctoral students. A more complete freedom arrives with tenure. However, by this stage of their career, the individual has invested nearly ten years of work in his original methodologies. As a respected and admired authority, one pressed by a cadre of attached doctoral students, and one on-track for a post as a senior professor, there is now more at risk in the prospect of a major philosophical and methodological shift. In the end, scholars arrive at the autumn of their careers, having spent 20 years trapped in an expedient and safe research framework originally selected "only" for their doctoral thesis.

One way to break this entrapment pattern of academic growth would be for thesis advisors to encourage more daring frameworks for postgraduate thesis work. Of course, those undertaking the higher-risk path must be made fully aware of the probabilities and impacts of the risks they are undertaking. However, encouragement toward higher-risk, higher-reward research, while involving risk at the early stages of an academic career, might inevitably lead to the benefit of relevance and diversity in the body of information systems research as a whole, while yielding more rewarding careers among IS scholars.

REFERENCES

Banville, C., & Landry, M. (1989). Can the field of MIS be disciplined? *Communications of the ACM, 32*, 48-61.

Baskerville, R. (1991). Risk analysis: An interpretive feasibility tool in justifying information systems security. *European Journal of Information Systems, 1*(2), 121-130.

Baskerville, R. (1993). Semantic database prototypes. *Journal of Information Systems, 3*(2), 119-144.

Baskerville, R. (1999). Investigating information systems with action research. *Communications of The Association for Information Systems, 19*(Article 2).

Baskerville, R., & Pries-Heje, J. (1999). Grounded action research: A method for understanding IT in practice. *Accounting, Management and Information Technology, 9,* 1-23.

Baskerville, R., & Wood-Harper, A. T. (1998). Diversity in information systems action research methods. *European Journal of Information Systems, 7*(2), 90-107.

Blum, F. (1955). Action research—a scientific approach? *Philosophy of Science, 22*(1), 1-7.

Checkland, P. (1981). *Systems thinking, systems practice.* New York: John Wiley & Sons.

Chiasson, M., & Dexter, A. S. (2001). System Development Conflict During the Use of an Information Systems Prototyping Method of Action Research: Implications for Practice and Research, *Information Technology and People,* 14 (March)

Clark, A. (1976). The client-practitioner relationship. In A. Clark (Ed.), *Experimenting with organizational life: The action research approach* (pp. 119-133). New York: Plenum.

Halfpenny, P. (1979). The analysis of qualitative data. *Sociological Review, 27,* 799-827.

Hult, M., & Lennung, S. (1980). Towards a definition of action research: A note and bibliography. *Journal of Management Studies, 17,* 241-250.

Keen, P. (1991). Relevance and rigor in information systems research: Improving quality, confidence cohesion and impact. In H.-E. Nissen & H. Klein & R. Hirschheim (Eds.), *Information systems research: Contemporary approaches & emergent traditions* (pp. 27-49). Amsterdam: North-Holland.

Kock, N., & McQueen, R. J. (1998). An action research study of effects of asynchronous groupware support on productivity and outcome quality of process redesign groups. *Journal of Organizational Computing and Electronic Commerce, 8*(2), 149-168.

Lau, F. (1997). A review on the use of action research in information systems studies. In A. Lee & J. Liebenau & J. DeGross (Eds.), *Information systems and qualitative research* (pp. 31-68). London: Chapman & Hall.

Lau, F., & Hayward, R. (2000). Building a virtual network in a community health research training program. *Journal of American Medical Informatics Association, 7,* 361-377.

Lewin, K. (1947a). Frontiers in group dynamics. *Human Relations, 1*(1), 5-41.

Lewin, K. (1947b). Frontiers in group dynamics II. *Human Relations, 1*(2), 143-153.

Mumford, E. (1983). *Designing human systems for new technology: The ethics method.* Manchester: Manchester: Manchester Business School.

Nosek, J. (1998). *Exploring it support for organizational learning in the virtual corporation* (Technical Report). Baltimore, Maryland: David D. Lattanze Center.

Orlikowski, W., & Baroudi, J. (1991). Studying information technology in organizations: Research approaches and assumptions. *Information Systems Research, 2*(1), 1-28.

Rapoport, R. (1970). Three dilemmas of action research. *Human Relations, 23*(6), 499-513.

Schein, E. (1987). *The clinical perspective in fieldwork.* Newbury Park, Calf: Sage.

Straub, D. W., & Welke, R. J. (1998). Coping with systems risk: Security planning models for management decision-making. *MIS Quarterly, 22*(4), 441-469.

Susman, G., & Evered, R. (1978). An assessment of the scientific merits of action research. *Administrative Science Quarterly, 23*(4), 582-603.

Thayer, H. S. (1981). Pragmatism: A reinterpretation of the origins and consequences. In R. Mulvaney & P. Zeltner (Eds.), *Pragmatism: Its sources and prospects* (pp. 1-20). Columbia: University of South Carolina Press.

Trist, E. (1976). Engaging with large-scale systems. In A. Clark (Ed.), *Experimenting with organizational life: The action research approach* (pp. 43-75). New York: Plenum.

Van Eynde, D., & Bledsoe, J. (1990). The changing practice of organization development. *Leadership & Organization Development Journal, 11*(2), 25-30.

Warmington, A. (1980). Action research: Its method and its implications. *Journal of Applied Systems Analysis, 7*(4), 23-39.

ENDNOTES

1 Two characteristics, a cyclical nature and an ethical framework, are excluded in this adaptation.

2 Further confirmation of an established and widely applied scientific theory might be a contribution, but only rarely considered a "significant" contribution.

3 The client did continue to build working applications on the design, but this was not the research objective and was not reported in the research paper.

4 Note that this study does not fit the profile of a "rising" risk setting because a particular event (the opportunity for action) changed the nature of the research. This is not a case where the risks of an action research study evolved, rather the risk profile appears *in situ* upon the introduction of action planning within the research project.

Chapter IX

A Classification Scheme for Interpretive Research in Information Systems

Heinz K. Klein
Temple University, USA

Michael D. Myers
University of Auckland, New Zealand

INTRODUCTION

Over the past decade tremendous progress has been made in the area of interpretive research in information systems. Whereas interpretive research was virtually non-existent within the IS research community at the beginning of the 1990s (Alavi & Carlson, 1992; Orlikowski & Baroudi, 1991), some ten years later interpretive research articles are regularly published in our conferences, journals, and books. The workshops, conferences (Cash & Lawrence, 1989; Lee, Liebenau, & DeGross, 1997; Mumford, Hirschheim, Fitzgerald, & Wood-Harper, 1985; Nissen, Klein, & Hirschheim, 1991), and special issues of journals devoted to qualitative and interpretive research in our field (Markus & Lee, 1999, 2000; Myers & Walsham, 1998) have had their intended effect, in that qualitative and interpretive research approaches are now accepted as part of the mainstream of the information systems research community (Markus, 1997).

Given the increase in the number of interpretive research articles being published in IS today, we believe it is timely to develop and explain a classification scheme of the literature. Such a classification scheme draws attention to the tremendous variety and breadth of interpretive research today, from the most abstract and general philosophical foundations to the most in-depth, detailed field studies. The explicit consideration of different types may contribute to a more effective division of labor among scholars with different research interests. It should also help interpretive researchers to better focus their work and to identify their research priorities.

Interpretive Research Defined

It is important to explicitly define what we mean by interpretive research. Following Myers (1997b) and Klein and Myers (1999), we make a clear distinction between qualitative and interpretive research. Qualitative research can be positivist, interpretive, or critical, depending upon the underlying philosophical assumptions of the researcher. The foundational assumption for interpretivists is that most of our knowledge is gained, or at least filtered, through social constructions such as language, consciousness, shared meanings, documents, tools, and other artifacts. Interpretive research does not predefine dependent and independent variables, but focuses on the complexity of human sense making as the situation emerges (Kaplan & Maxwell, 1994); it attempts to understand phenomena through the meanings that people assign to them (Boland, 1985, 1991; Orlikowski & Baroudi, 1991). This has implications for the types of theories and methods upon which interpretivists can draw. Interpretive methods of research in IS are "aimed at producing an understanding of the context of the information system, and the process whereby the information system influences and is influenced by the context" (Walsham, 1993, pp. 4-5). In summary, interpretive theories, methods, empirical investigations and interventions differ from their positivist counterparts in their epistemological premises. For a more detailed discussion of these terms we refer the reader elsewhere (Klein & Myers, 1999; Myers, 1997b; Orlikowski & Baroudi, 1991).

The Scope and Approach of This Chapter

The purpose of this chapter is to suggest a classification scheme for interpretive research. It defines five types of interpretive research and presents representative examples for them. These five types can be grouped into general foundations of interpretive research (the first three categories), and applications of interpretive research to information systems (the last two categories). The first three categories apply to interpretive research in general, because interpretive researchers in all disciplines rely on certain philosophical foundations, theories and methods. The last two categories are IS specific in the sense that they apply to theoretical concepts and empirical phenomena in the field of information systems.

Our intent in providing such a classification is to testify to the breadth of interpretive research today and to give some examples of the different types of interpretive literature. Of course, we do not expect that every interpretive article will neatly fit in one particular category. Whereas most publications are likely to make their principal contribution to only one category, some might contribute equally to more than one. In particular, the pioneering works of interpretive research in information systems had to contribute to several frontiers at once in order to be comprehensible and credible to a wider audience. Therefore it should be expected that they will be referenced in more than one category. Also, it is possible that new kinds of interpretive research may emerge, which may require a broadening of the classification scheme in the future.

This chapter is organized as follows. The next section defines five principal types or categories of interpretive research. It also applies the typology to examples of the existing literature to improve our understanding of which areas are well understood and which have received insufficient attention. The following section discusses some of the important implications of our proposed framework. The conclusions argue that the acceptance of typologies as an organizing construct by the research community at large is an important prerequisite for building a cumulative tradition in interpretive research. Classification of prior work is an important part of every science.

THE PRINCIPAL CATEGORIES
OF INTERPRETIVE RESEARCH

Table 1 summarizes the five principal types of interpretive research, which we have identified from a survey of the published literature. It also gives at least one example for each category. Each category is discussed in more detail below.

Establishing the Philosophical
Foundations of Interpretive Research

The work in this category constructs the philosophical and conceptual foundations of interpretivism. Professional philosophers define the most fundamental assumptions of interpretivism, especially its epistemological and ethical premises. They have contributed at least two different lines of philosophical thinking to the foundations of interpretive research.

One school of thought focuses on human intentions in the use of language and various methods for understanding the meaning of language (such as speech act theory, conversation and discourse analysis). The intellectual roots of this line of research can be traced to Wittgenstein's ideas on meaning and language (Wittgenstein, 1958) and its subsequent refinements (e.g., Austin, 1962; Searle, 1979).

The second school of thought has maintained close links to phenomenology and hermeneutics, and focuses on subjective consciousness, i.e., the general conditions of being human and meaning ("being" in the sense of Heidegger). The intellectual roots of this line of research are Husserl (1970, 1982), Heidegger (1962), Gadamer (1975, 1976) and Ricoeur (1974,1981).

Although the scholarly work in this category is primarily the domain of professional philosophers, IS researchers typically draw upon this work to build research frameworks, theories and methods related to information systems. For example, the principle of the hermeneutic circle, which is discussed and explained by interpretive philosophers, is fundamental to all interpretive research of a hermeneutic nature, including that in IS. As another example, Schultze (2000) cites a philosophical definition of practice-oriented research (Pickering, 1992; Turner, 1994) in the context of explaining her research method for a confessional account of knowledge work.

Table 1: Principal Categories of Interpretive Research

FOUNDATIONS OF INTERPRETIVE RESEARCH	**1. Establishing the Philosophical Foundations of Interpretive Research** Constructs the conceptual foundations of interpretivism, e.g. by defining its most fundamental assumptions such as its epistemological and ethical premises. Mostly pursued by philosophers. Examples: (Gadamer, 1975, 1976; Husserl, 1970; Husserl, 1982; Ricoeur, 1974, 1981).
	2. Building Interpretive Social Theories Uses the literature on the philosophical foundations of interpretivism to create theories about social phenomena. The influence is often eclectic in that social thinkers in their theories typically try to bridge more than one philosophical tradition. Their influence may not always be explicit, but often can be discerned in bibliographical citations. Mostly pursued by social scientists. Example: Giddens (1984) proposed structuration theory as an integrative perspective which recognizes both subjective and objective dimensions of social reality.
	3. Advancing Interpretive Research Methods Clarifies the nature of interpretive research methods, formulates methodological standards, or advises others how to properly employ interpretive methods. This kind of work tends to draw directly on the philosophical foundations of interpretive research and the scholarly discussion in the philosophy of science. In principle these methodological advances could apply to any discipline with an interpretive research interest, but here we focus only on the work done by information systems researchers. Example: Boland (1985) explained hermeneutics and phenomenology for an audience with applied research interests and established their relevance for IS research. This paper was a contribution to advancing interpretive research methods in IS.
APPLICATIONS OF INTERPRETIVE RESEARCH TO IS	**4. Applying Interpretive Concepts to Advance our Understanding of an IS Research Area** Applies and adapts theories and concepts from categories 1 and 2 to advance our understanding of a substantive area or identifiable body of knowledge of IS research. For example, interpretive theories and concepts may help to (re)define a specialization (such as groupware or computer supported cooperative work), critique the thrust of prior work in IS, or point out new directions for IS research. Examples: Lyytinen and Ngwenyama (1992) used structuration theory to coin a new definition of the nature of CSCW. Lee (1994) used hermeneutics to theorize about email use and to critique prior work in IS which uses information richness theory.
	5. Applying an Interpretive Frame of Reference in Empirical Investigations and Interventions Applies the concepts and methods from any or all of the above to guide field studies (e.g. interpretive case studies, ethnographies), interpretive experiments, or practical interventions (e.g. action research) within the domain of information systems. Examples: Orlikowski (1991) used the ethnographic research method to study a large, multinational software consulting firm. Bentley et al. (1992) used the ethnographic method to gain an understanding of human cooperation in air traffic control. Ytterstad et al.'s (1996) action research project developed a communication system for local politicians.

Building Interpretive Social Theories

The research in this category uses the literature on the philosophical foundations of interpretivism to create theories about social phenomena. The influence is often eclectic in that the originators of social theories typically try to bridge more than one philosophical thinker or philosophical tradition. The multiple sources of philosophical inspiration may not always be explicit, but often can be discerned in bibliographical citations. Although research in this category is primarily the domain of historians or social scientists, typically IS researchers draw upon this work to build research frameworks and theories related to information systems.

For example, Giddens (1984), a British sociologist, proposed structuration theory, in which social reality is seen as having a duality of structure: social structure constrains human actions, but at the same time every social action incrementally changes the structure. Giddens' theory has been taken up by many IS researchers (e.g. Orlikowski, 1992; Orlikowski & Robey, 1991) and used (usually with some modifications) to explain IS phenomena. Many other social theories with interpretive components have also influenced IS research. For example, Weber (1930,1949) focused on the interpretive understanding of human action, and saw social systems as the result of meaningful human action. Expanding on Weber, Berger and Luckman (1967), and Schutz (1982) clarified how humans create a socially shared reality through processes of reification. Blumer's (1969) symbolic interactionism, Geertz's (1973, 1983) cultural anthropology, and Latour's (1993, 1996a, 1996b) actor-network theory are among some of the other social theories that have been widely used in IS research.

Advancing Interpretive Research Methods

The research in this category clarifies the nature of interpretive research methods, formulates methodological standards, or advises others how to properly employ interpretive methods. It tends to draw directly on the philosophical foundations of interpretive research and the scholarly discussion in the philosophy of science. Besides the more general methodological literature in the social sciences, each discipline also tends to have its own discourse on research methods. Here we focus only on the work done by information systems researchers.

Boland and Wynn were the first to introduce interpretive research methods into the field of information systems during the late 1970s. Boland pioneered the hermeneutic-phenomenological line of research (Boland, 1978,1979; Boland & Day, 1989). He not only explained hermeneutics and phenomenology for an audience with applied research interests and established their relevance for IS research (Boland, 1985), but also contributed to advancing interpretive research methods in IS by performing an interpretive experiment.

In interpretive experiments, the data is gathered from contrived (or artificially created) situations, whereas in field studies the data is gathered from natural (or rather, social) settings. Although very few interpretive researchers have used experiments—probably because experiments are typically associated with positivist research—we believe that interpretivism and experiments are not necessarily incompatible. The interpretive experiments differ from positivist studies in their epistemological premises. In positivist experiments, the researcher assumes that he or she can learn about the phenomena under investigation by building a series of models consisting of relationships between dependent and independent variables. In each experimental cycle the independent variables are set to different values and the outcome is measured and compared to the predicted behavior of the dependent variables. If the prediction fails, the model is modified. In interpretive experiments, the researcher assumes that the subjects' behavior is best explained by the meanings that they assign to the cues given to them in the experiment. The experimenter tries to understand the reasons why people acted in the ways observed by focusing on the meanings that they constructed in making sense of the experimental situation. There are no a priori models and it is not assumed that the phenomena under investigation can be separated into dependent and independent variables. As early as 1979, Boland (1979) set up an experiment to study the sense-making capacities of accountants when giving them contradictory financial reports. The experiential accounts collected in the experiments were used to demonstrate the usefulness of phenomenological concepts for understanding how accountants make sense of financial reports.

Wynn's study (1979) of the behavior of clerks handling customer complaints broke the conceptual ground of introducing language analysis methods into IS research. Although this dissertation was never published, it influenced several IS research projects. Later, language analysis methods were substantially broadened by the introduction of semiotic concepts (Andersen, 1991; Andersen & Holmqvist, 1995; Holmqvist & Andersen, 1987), speech act theory and social action typologies (Auramaki, Lehtinen, & Lyytinen, 1988; Dietz & Widdershoven, 1991; Klein & Truex III, 1995).

Since the early 1990s, interpretive field studies have been the most common interpretive research method for conducting interpretive empirical investigations. This kind of work depends on gaining access to people in the field. One of the first such studies was Suchman (1987). Following our earlier work (Klein & Myers, 1999), we define an interpretive field study to include interpretive in-depth case studies (Walsham, 1995) and ethnographies (Bentley et al., 1992; Harvey, 1997; Harvey & Myers, 1995; Myers, 1997a; Wynn, 1991; Zuboff, 1988). Although there is no hard and fast distinction, the principal difference between the two depends upon the length of time that the investigator is required to spend in the field and the extent to which the researcher immerses himself or herself in the life of the social group under study. As Yin (1994, pp. 10-11) describes, "Ethnographies usually require long periods of time in the 'field' and emphasize detailed, observational evidence . . . In contrast, case studies are a form of enquiry that does *not* depend solely on ethnographic or participant-observer data."

So far we have introduced four types of interpretive research methods: interpretive experiments, language analysis, in-depth case studies, and ethnographies. In addition, a variety of other interpretive research methods have been suggested for IS research. Amongst those that have been discussed are Idhe's phenomenology (Rathswohl, 1991), deconstruction (Beath & Orlikowski, 1994), action research (Avison, Lau, Myers, & Nielson, 1999; Baskerville, 1999; Baskerville & Wood-Harper, 1996; Lau, 1997), and grounded theory (Trauth & Jessup, 2000; Urquhart, 1997). In addition we would like to mention ethnomethodology (Garfinkel, 1967) as another possible research method that could be applied to IS.

Both action research and grounded theory are particularly interesting methods because, in a similar way to case studies, they can be applied in both positivist and interpretive research. An introduction to the different types of action research methods (positivist, interpretive, and critical) is given in Myers (1997b) and to other types by Baskerville and Wood-Harper (1998). In the original formulation of grounded theory (Glaser & Strauss, 1967), it was supposed to allow categories to emerge "naturally" from the data – which is, of course, a positivist notion. Yet grounded theory also incorporates a number of important techniques that are well suited to supporting the hermeneutic circle, e.g., open and axial coding. It is beyond the scope of this chapter to examine the methodological controversies surrounding action research and grounded theory, therefore in the rest of this chapter will only refer to interpretive uses of these methods.

The question of how to use various qualitative and interpretive research methods has been taken up in various conferences (e.g. Cash & Lawrence, 1989; Lee et al., 1997; Mumford et al., 1985; Nissen et al., 1991). A special issue of the *Journal of Information Technology* was devoted to interpretive research in 1998 and included some methods papers (Myers & Walsham, 1998). More recently, a special issue of *MIS Quarterly* was devoted to "intensive research" in information systems (Markus & Lee, 1999, 2000). This special issue included methods articles by Klein and Myers (1999), who suggested a set of principles for conducting and evaluating interpretive field studies in information systems and Schultze (2000), who suggested how a confessional ethnography can be evaluated. In the same special issue, the article by Trauth and Jessup (2000) also contributes to the methods discussion by applying a grounded theory interpretive analysis to group support systems data. The data were collected in a positivist research project and consisted of discussion transcripts and contextual data relating to the organization. The article illustrates the use of multiple research methods in the same study and derives four criteria for evaluating their results (triangulation, authenticity, breakdown resolution, and replication).

A very interesting, but also controversial methodological contribution has been made by Lee's insistent claim that interpretive methods can bridge more than one research approach or perspective. Lee (1991) gave an example of how interpretivist and positivist research methods can be combined. Ngwenyama and Lee (1997) suggested that Critical Social Theory (CST) incorporates and offers additional insights over the interpretivist perspective. In fact they go so far as to assert that CST overcomes the weakness of the positivist and interpretivist perspectives, at least in relation to their development of a new theory of communication richness in electronic media. Clearly, such methodological claims call for further analysis, because they challenge not only Burrell and Morgan's (1979, p.398) incommensurability conjecture (paradigms "stand as four mutually exclusive ways of seeing the world"), but the very independence of paradigms. (Can one paradigm compensate for the weakness of another or is it even possible for one paradigm to simply import others and limit their scope to certain ontological subdomains, as CST appears to suggest?) While further elaboration of these issues lies beyond the scope of this chapter, their emergence in premier journals is vivid testimony that research on interpretive research methods is currently a topic of great interest.

Applying Interpretive Concepts to Advance Our Understanding of an IS Research Area

The research in this category adapts and applies theories and concepts from philosophy (Category 1) and/ or social theory (category 2) to a substantive area or given body of knowledge of IS research. For example IS researchers have written articles on defining a specialization, such as groupware or computer supported cooperative work (cf. Lyytinen and Ngwenyama, 1992), on critiquing the thrust of prior work in IS, or pointing out new directions for future IS research. Such work is useful because it advances our understanding of the IS field.

Probably the first contribution in this category was by Boland (1978; 1979). He confronted the rational decision-making viewpoint of the nature of IS use, prevailing at that time, with an interpretivist critique. Boland pointed out that the key role of an information system may not be to supply data or automate decision making, but to intervene in perceptions and language by which people make sense of their organizational environment and construct a socially shared reality. The argumentation was based on Blumer's (1969) symbolic interactionism. In later work, Boland (1985) explained and interpreted phenomenology for IS researchers. This was the first attempt to bring one of the most important lines of philosophical thinking to the IS research community.

Other important examples in this category are analyses of how the philosophy and theory of language apply to IS research. Wynn's (1979) research on conversation analysis is one of the first examples of language analysis opening up an new avenue of thinking in IS. The variety of more recent research orientations and methods in language analysis becomes obvious if Wynn's approach to conversation analysis is compared with some recent work in semiotic analysis, e.g., Andersen (1991), Forester (1992) or Andersen and Holmqvist (1995). The latter make more rigid assumptions about a coding scheme to tabulate different "language moves" than Wynn.

The most detailed theoretical constructions which have emerged in this category come from the application of speech act theory (Searle, 1979). One is the Language Action theory of Lehtinen and Lyytinen (1986). Their theory views an IS as an artifact which replaces more conventional forms of communication (such as telephone calls or meetings). Lehtinen and Lyytinen provided a formalism to model the meanings which need to be captured by an IS so that there is no loss of meanings through the use of IS and the integrity of the human communication is maintained. They insisted that this is important for determining the requirements for an IS. Also based on speech-act theory, Winograd and Flores (1987) proposed the Coordinator, which prompts its users to follow through on the commitments made in prior interactions. Another example of the application of language analysis philosophy to IS research is Dietz and Widdershoven's (1991) comparison of Searle's and Habermas' (1984) speech act classifications. They concluded that the latter is the more useful starting point for information systems research, in particular for information systems modeling and requirements analysis.

There are many other examples of the application of interpretive concepts from philosophy or social theory to information systems research. For instance, Orlikowski and Robey (1991) proposed structuration theory as a way of studying the relationship between information systems and organizations; Lyytinen and Ngwenyama (1992) used structuration theory to coin a new definition of the nature of CSCW as was already mentioned; Lee (1994) used hermeneutics to critique prior work in IS, which uses information richness theory; Myers (1994) used critical hermeneutics to critique prior work on the implementation of information systems; and Walsham (1997) assessed the current and potential future contribution of actor-network theory to IS research.

Applying an Interpretive Frame of Reference in Empirical Investigations and Interventions

The work in this category uses the concepts and methods from any or all of the categories above to guide empirical investigations (e.g. interpretive case studies or ethnographies) or practical interventions (e.g. action research). As we suggested earlier, there exist at least four generally recognized options for conducting empirical investigations using an interpretive frame of reference; these are interpretive experiments, language analysis methods, in-depth case studies and ethnographies.

Surprisingly few *interpretive experiments* have been published. The only one that has come to our attention since Boland (1979) has been Klein and Hirschheim's (1983) scenario study to assess the likely impact of new technologies on different organizational stakeholder groups. They proposed to use a scenario simulation in which different stakeholder representatives would act out their reactions to intended IT changes. (In 1983 the IT change in question was office automation, but the method is applicable to many other types of changes as long as they can be prototyped in some form.) They then showed how the results of such a social simulation could be interpretively analyzed to gain substantive insights about the likely reactions of stakeholder groups. This kind of study is like a military maneuver that can be used to anticipate problems and likely human reactions by all concerned.

Some early examples of empirical investigations using *language analysis methods* are Wynn (1991), Klein and Truex (1995) and Holmqvist and Andersen (1987).

A detailed analysis of two in-depth case studies and one ethnography can be found in Klein and Myers (1999, pp. 79-86). Other examples of interpretive in-depth case studies are Barrett and Walsham (1999), Bussen and Myers (1997), Myers (1994), Walsham and Waema (1994), and Walsham and Sahay (1999). A few examples of interpretive ethnographies are Komito (1998), Orlikowski (1991), Sayer (1998) and Schultze (2000). Two examples of interpretive action research are Olesen and Myers (1999) and Ytterstad et al. (1996). In the first study, the authors used structuration theory as the theoretical lens for interpreting the data collected through unstructured interviews, documents, and participant observation. Throughout the action research cycle, from diagnosing through to evaluating and specifying learning (Susman & Evered, 1978), these data were used in a way that is consistent with interpretivist epistemology. At the end the authors constructed a story to make sense of the changes in work habits and organizational structure which occurred as a result of implementing a groupware product. The Ytterstad et al. (1996) example relates to Norway's Telenor Research and Development, which developed and implemented a communication system to support local politicians. The system provided a graphical interface to enable politicians to make phone calls, set up teleconferences and exchange documents. A two-year field trial, involving 35 municipal politicians and their administration, demonstrated its value. As this work is published in *MISQ Discovery* using the capabilities of the World Wide Web, this article is an excellent source of material if one wishes to look in depth at the distinguishing features of an interpretive action research project.

Many more interpretive empirical studies have appeared in the books published under the auspices of the IFIP Working Group 8.2 (see e.g. Avison, Kendall, & DeGross, 1993; Baskerville, Smithson, Ngwenyama, & Degross, 1994; Larsen, Levine, & DeGross, 1998; Lee et al., 1997; Ngwenyama, Introna, Myers, & DeGross, 1999; Orlikowski, Walsham, Jones, & DeGross, 1996).

DISCUSSION AND CONCLUSIONS

This chapter has suggested a classification scheme for interpretive research. It has identified five different types of interpretive research (Table 1), grouped into general foundations of interpretive research (the first three categories), and applications of interpretive research to information systems (the last two categories). We have also given representative examples of each category. The classification scheme draws attention to the tremendous variety and breadth of interpretive research today, from the most abstract and general philosophical foundations to the most concrete and detailed field studies. We believe that our proposed classification scheme is a useful contribution to IS research, leaving open the question of whether it might be applicable to other disciplines as well.

From an IS perspective, we see at least two major contributions. First, it might help IS researchers—especially those who are starting an empirical interpretive study or writing a paper for publication—to identify and characterize their most important contribution. For example, interpretive researchers might ask themselves the question: is this paper primarily a contribution to IS research methods, is it primarily a contribution to advancing our understanding of an IS research area, or is it primarily an empirical investigation? Although it is possible for articles to be in multiple categories (since we are not suggesting that the categories are mutually exclusive), the primary contribution of a paper is most likely to be in just one category. Identifying the particular category for a specific project should help interpretive researchers to better focus their work.

Second, the proposed classification scheme helps to identify how those studies in the category of empirical investigations can be built on one or more of the other four. Generally speaking, we believe that every interpretive empirical investigation in IS (Category 5), as a general rule, should build primarily on research work in Categories 2-4. That is, every interpretive researcher needs guidance with regard to the use of appropriate research methods, and the empirical work needs to be guided by (or at least informed by) one or more social theories (such as structuration theory). We would expect every interpretive paper reporting on an empirical investigation to have a discussion (and an appropriate literature review) of both the interpretive methods employed, the theory, and how this theory has been applied in prior IS research. If this is true, then this chapter helps IS researchers to identify the categories of work upon which they should build in the light of their own particular research objectives.

Our view, therefore, is that the IS discipline will start to develop a cumulative tradition (if it has not already started to do so) as more IS research articles are published in Categories 3-5. Our proposed classification scheme makes the interconnections and mutual dependencies between the different categories of IS research more visible. A further implication of this is that a more effective division of labor may emerge amongst IS researchers. Some IS scholars are more interested in theoretical and conceptual development, some are more interested in methodological issues, whereas others prefer to get involved with empirical investigations. In the future, perhaps the only ones who will go back to the philosophical or sociological literature will be those IS scholars who want to generate new conceptual frameworks or research methods. These researchers will ensure that there is continual theoretical development within the field informing those interpretive IS researchers who prefer to focus on empirical investigations.

Of equal importance for the success of the theoretical work is that those who specialize in it keep abreast of the advances that come from the work in Category 5. Theory needs to be confronted with the results from empirical studies, but this is only a first step. The ultimate goal must be to use the empirical studies as a base to construct higher-order theories that organize and explain the findings of as many studies as possible. At this point we do not know whether this will lead to ever broader theoretical frameworks which can explain an ever larger number of studies or, alternatively, to various competing theoretical generalizations of limited scope. It seems to us that this difficult, but important work has barely begun. In any case we can see that both theoretical advances and empirical discoveries depend upon each other if any progress is to be made in the field. Our classification scheme highlights the importance of theoretical advances, methodological development, and empirical studies, and how work in all three is needed for the advancement of interpretivism in information systems research.

REFERENCES

Alavi, M., & Carlson, P. (1992). A review of MIS research and disciplinary development. *Journal of Management Information Systems, 8*(4), 45-62.

Andersen, P. B. (1991). A Semiotic Approach to Construction and Assessment of Computer Systems. In H.-E. Nissen, H. K. Klein, & R. A. Hirschheim (Eds.), *Information Systems Research: Contemporary Approaches and Emergent Traditions* (pp. 465-514). Amsterdam: North-Holland.

Andersen, P. B., & Holmqvist, B. (Eds.). (1995). *The Semiotics of the Workplace*. Berlin: Walter De Gruyter.

Auramaki, E., Lehtinen, E., & Lyytinen, K. (1988). A Speech-Act Based Office Modeling Approach. *ACM Transactions on Office Information Systems, 6*(2), 126-152.

Austin, J. (1962). *How to do things with words*. Oxford: Clarendon Press.

Avison, D., Lau, F., Myers, M. D., & Nielson, P. A. (1999). Action Research. *Communications of the ACM, 42*(1), 94-97.

Avison, D. E., Kendall, J. E., & DeGross, J. I. (Eds.). (1993). *Human, Organizational, and Social Dimensions of Information Systems Development*. Amsterdam: North Holland.

Barrett, M., & Walsham, G. (1999). Electronic Trading and Work Transformation in the London Insurance Market. *Information Systems Research, 10*(1), 1-22.

Baskerville, R. (1999). Investigating Information Systems with Action Research. *Communications of the AIS, 2*(19), http://cais.isworld.org/

Baskerville, R., Smithson, S., Ngwenyama, O., & Degross, J. I. (Eds.). (1994). *Transforming Organizations with Information Technology*. Amsterdam: North-Holland.

Baskerville, R. L., & Wood-Harper, A. T. (1996). A Critical Perspective on Action Research as Method for Information Systems Research.

Journal of Information Technology, 11, 235-246.

Baskerville, R. L., & Wood-Harper, A. T. (1998). Diversity in information systems action research methods. *European Journal of Information Systems, 7*, 90-107.

Beath, C. M., & Orlikowski, W. J. (1994). The Contradictory Structure of Systems Development Methodologies: Deconstructing the IS-User Relationship in Information Engineering. *Information Systems Research, 5*(4), 350-377.

Bentley, R., Hughes, J. A., Randall, D., Rodden, T., Sawyer, P., Shapiro, D., & Sommerville, I. (1992). *Ethnographically-informed systems design for air traffic control.* Paper presented at the ACM 1992 Conference on Computer-Supported Cooperative Work: Sharing Perspectives, New York.

Berger, P., & Luckmann, T. (1967). *The Social Construction of Reality.* Middlesex: Penguin.

Blumer, H. (1969). *Symbolic interactionism; perspective and method.* Englewood Cliffs, NJ: Prentice Hall.

Boland, R. (1978). The process and product of system design. *Management Science, 28*(9), 887-898.

Boland, R. (1979). Control, causality and information system requirements. *Accounting, Organizations and Society, 4*(4), 259-272.

Boland, R. (1985). Phenomenology: A Preferred Approach to Research in Information Systems. In E. Mumford, R. A. Hirschheim, G. Fitzgerald, & A. T. Wood-Harper (Eds.), *Research Methods in Information Systems* (pp. 193-201). Amsterdam: North Holland.

Boland, R. J. (1991). Information System Use as a Hermeneutic Process. In H.-E. Nissen, H. K. Klein, & R. A. Hirschheim (Eds.), *Information Systems Research: Contemporary Approaches and Emergent Traditions* (pp. 439-464). Amsterdam: North-Holland.

Boland, R. J., & Day, W. F. (1989). The Experience of System Design: A Hermeneutic of Organizational Action. *Scandinavian Journal of Management, 5*(2), 87-104.

Burrell, G., & Morgan, G. (1979). *Sociological Paradigms and Organizational Analysis.* Portsmouth, NH: Heinemann.

Bussen, W., & Myers, M. D. (1997). Executive Information Systems Failure: A New Zealand Case Study. *Journal of Information Technology, 12*(2), 145-153.

Cash, J. I., Jr., & Lawrence, P. R. (Eds.). (1989). *Qualitative Research Methods.* (Vol. 1). Boston, MA: Harvard Business School.

Dietz, J. L. G., & Widdershoven, G. A. M. (1991). *Speech Acts or Communicative Action?* Paper presented at the Second European Conference on Computer Supported Cooperative Work, Dordrecht.

Forester, J. (1992). Critical ethnography: on field work in an Habermasian way. In M. Alvesson & H. Willmott (Eds.), *Critical Management Studies* (pp. 46-65). London: Sage Publications.

Gadamer, H.-G. (1975). *Truth and Method.* New York: Seasbury Press.

Gadamer, H.-G. (1976). *Philosophical Hermeneutics.* California: University of California Press.

Garfinkel, H. (1967). *Studies in ethnomethodology.* Englewood Cliffs, NJ: Prentice-Hall.

Geertz, C. (1973). *The Interpretation of Cultures.* New York: Basic Books.

Geertz, C. (1983). *Local Knowledge: Further Essays in Interpretive Anthropology.* New York: Basic Books.

Giddens, A. (1984). *The Constitution of Society: Outline of a Theory of Structure.* Berkeley: University of California Press.

Glaser, B. G., & Strauss, A. (1967). *The Discovery of Grounded Theory: Strategies for Qualitative Research.* Chicago: Aldine Publishing Company.

Habermas, J. (1984). *The Theory of Communicative Action.* (Vol. 1). Boston: Beacon Press.

Harvey, L. (1997). A Discourse on Ethnography. In A. S. Lee, J. Liebenau, & J. I. DeGross (Eds.), *Information Systems and Qualitative Research* (pp. 207-224). London: Chapman and Hall.

Harvey, L., & Myers, M. D. (1995). Scholarship and practice: the contribution of ethnographic research methods to bridging the gap. *Information Technology & People, 8*(3), 13-27.

Heidegger, M. (1962). *Being and Time.* Oxford: Basil Blackwell.

Holmqvist, B., & Andersen, P. B. (1987). Work-language and information technology. *Journal of Pragmatics, 11*, 327-357.

Husserl, E. (1970). *Logical Investigations* (Findlay, J N, Trans.). London: Routledge and Kegan Paul.

Husserl, E. (1982). *Ideas pertaining to a pure phenomenology and to a phenomenological philosophy.* Boston: Kluwer.

Kaplan, B., & Maxwell, J. A. (1994). Qualitative Research Methods for Evaluating Computer Information Systems. In J. G. Anderson, C. E. Aydin, & S. J. Jay (Eds.), *Evaluating Health Care Information Systems: Methods and Applications* (pp. 45-68). Thousands Oaks:

Sage.

Klein, H. K., & Hirschheim, R. (1983). Issues and Approaches to Appraising Technological Change in the Office: A Consequentialist Perspective. *Office: Technology and People, 2*, 15-24.

Klein, H. K., & Myers, M. D. (1999). A Set of Principles for Conducting and Evaluating Interpretive Field Studies in Information Systems. *MIS Quarterly, 23*(1), 67-93.

Klein, H. K., & Truex III, D. P. (1995). Discourse Analysis: A Semiotic Approach to the Investigation of Organizational Emergence. In P. B. Andersen & B. Holmqvist (Eds.), *The Semiotics of the Workplace* (pp. 227-268). Berlin: Walter De Gruyter.

Komito, L. (1998). Paper 'work' and electronic files: defending professional practice. *Journal of Information Technology, 13*(4), 235-246.

Larsen, T. J., Levine, L., & DeGross, J. I. (Eds.). (1998). *Information Systems: Current Issues and Future Changes*. Laxenburg: International Federation of Information Processing.

Latour, B. (1993). *We have never been modern*. Hemel Hempstead: Harvester Wheatsheaf.

Latour, B. (1996a). *Aramis or the Love of Technology*. Cambridge, MA: Harvard University Press.

Latour, B. (1996b). Social Theory and the Study of Computerized Work Sites. In W. Orlikowski, G. Walsham, M. R. Jones, & J. I. DeGross (Eds.), *Information Technology and Changes in Organizational Work* (pp. 295-307). London: Chapman and Hall.

Lau, F. (1997). A Review on the Use of Action Research in Information Systems Studies. In A. S. Lee, J. Liebenau, & J. I. DeGross (Eds.), *Information Systems and Qualitative Research* (pp. 31-68). London: Chapman and Hall.

Lee, A. S. (1991). Integrating Positivist and Interpretive Approaches to Organizational Research. *Organization Science, 2*(4), 342-365.

Lee, A. S. (1994). Electronic mail as a medium for rich communication: An empirical investigation using hermeneutic interpretation. *MIS Quarterly, 18*(2), 143-157.

Lee, A. S., Liebenau, J., & DeGross, J. I. (Eds.). (1997). *Information Systems and Qualitative Research*. London: Chapman and Hall.

Lehtinen, E., & Lyytinen, K. (1986). Action Based Model of Information Systems. *Information Systems, 13*(4), 299-317.

Lyytinen, K. J., & Ngwenyama, O. K. (1992). What does Computer Support for Cooperative Work Mean? A Structurational Analysis of

Computer Supported Cooperative Work. *Accounting, Management and Information Technologies, 2*(1), 19-37.

Markus, M. L. (1997). The Qualitative Difference in Information Systems Research and Practice. In A. S. Lee, J. Liebenau, & J. I. DeGross (Eds.), *Information Systems and Qualitative Research* (pp. 11-27). London: Chapman and Hall.

Markus, M. L., & Lee, A. S. (1999). Special Issue on Intensive Research in Information Systems: Using Qualitative, Interpretive, and Case Methods to Study Information Technology - Foreward. *MIS Quarterly, 23*(1), 35-38.

Markus, M. L., & Lee, A. S. (2000). Special Issue on Intensive Research in Information Systems: Using Qualitative, Interpretive, and Case Methods to Study Information Technology - Second Installment - Foreward. *MIS Quarterly, 24*(1), 1-2.

Mumford, E., Hirschheim, R. A., Fitzgerald, G., & Wood-Harper, A. T. (Eds.). (1985). *Research Methods in Information Systems.* New York: North-Holland.

Myers, M. D. (1994). A disaster for everyone to see: An interpretive analysis of a failed IS project. *Accounting, Management and Information Technologies, 4*(4), 185-201.

Myers, M. D. (1997a). Critical Ethnography in Information Systems. In A. S. Lee, J. Liebenau, & J. I. DeGross (Eds.), *Information Systems and Qualitative Research* (pp. 276-300). London: Chapman and Hall.

Myers, M. D. (1997b). Qualitative Research in Information Systems. *MIS Quarterly, 21*(2), 241-242. *MISQ Discovery*, archival version, June 1997, www.misq.org/misqd961/isworld/. *MISQ Discovery*, updated version, September 2000, www.auckland.ac.nz/msis/isworld/

Myers, M. D., & Walsham, G. (1998). Guest Editorial: Exemplifying Interpretive Research in Information Systems: An Overview. *Journal of Information Technology, 13*(4), 233-234.

Ngwenyama, O. K., Introna, L. D., Myers, M. D., & DeGross, J. I. (Eds.). (1999). *New Information Technologies in Organizational Processes: Field Studies and Theoretical Reflections on the Future of Work.* Norwell, MA: Kluwer Academic Publishers.

Ngwenyama, O. K., & Lee, A. S. (1997). Communication Richness in Electronic Mail: Critical Social Theory and the Contextuality of Meaning. *MIS Quarterly, 21*(2), 145-167.

Nissen, H.-E., Klein, H. K., & Hirschheim, R. A. (Eds.). (1991).

Information Systems Research: Contemporary Approaches and Emergent Traditions. Amsterdam: North-Holland.

Olesen, K., & Myers, M. D. (1999). Trying To Improve Communication And Collaboration With Information Technology: An Action Research Project Which Failed. *Information Technology & People, 12*(4), 317-332.

Orlikowski, W. J. (1991). Integrated Information Environment or Matrix of Control? The Contradictory Implications of Information Technology. *Accounting, Management and Information Technologies, 1*(1), 9-42.

Orlikowski, W. J. (1992). The Duality of Technology: Rethinking the Concept of Technology in Organizations. *Organization Science, 3*(3), 398-427.

Orlikowski, W. J., & Baroudi, J. J. (1991). Studying Information Technology in Organizations: Research Approaches and Assumptions. *Information Systems Research, 2*(1), 1-28.

Orlikowski, W. J., & Robey, D. (1991). Information Technology and the Structuring of Organizations. *Information Systems Research, 2*(2), 143-169.

Orlikowski, W. J., Walsham, G., Jones, M. R., & DeGross, J. I. (Eds.). (1996). *Information Technology and Changes in Organizational Work.* London: Chapman and Hall.

Pickering, A. (1992). From Science as Knowledge to Science as Practice. In A. Pickering (Ed.), *Science as Practice and Culture* (pp. 1-26). Chicago: University of Chicago Press.

Rathswohl, E. J. (1991). Applying Don Idhe's phenomenology of instrumentation as a framework for designing research in information science. In H. E. Nissen, H. K. Klein, & R. A. Hirschheim (Eds.), *Information Systems Research: Contemporary Approaches and Emergent Traditions* (pp. 421-438). Amsterdam: North-Holland.

Ricoeur, P. (1974). *The Conflict of Interpretations: Essays in Hermeneutics.* Evanston: Northwestern University Press.

Ricoeur, P. (1981). *Hermeneutics and the Human Sciences.* Cambridge: Cambridge University Press.

Sayer, K. (1998). Denying the technology: Middle management resistance in business process re-engineering. *Journal of Information Technology, 13*(4), 247-257.

Schultze, U. (2000). A Confessional Account of an Ethnography about Knowledge Work. *MIS Quarterly, 24*(1), 3-41.

Schutz, A. (1982). *Life forms and meaning structure* (Wagner, Helmut R, Trans.). London: Routledge and Kegan Paul.

Searle, J. (1979). *Expression and Meaning*. Cambridge: Cambridge University Press.

Suchman, L. (1987). *Plans and Situated Actions: The Problem of Human-Machine Communication*. Cambridge: Cambridge University Press.

Susman, G. I., & Evered, R. D. (1978). An Assessment of the Scientific Merits of Action Research. *Administrative Science Quarterly, 23*(4), 582-603.

Trauth, E. M., & Jessup, L. M. (2000). Understanding Computer-Mediated Discussions: Positivist and Interpretive Analyses of Group Support System Use. *MIS Quarterly, 24*(1), 43-79.

Turner, S. (1994). *The Social Theory of Practices: Tradition, Tacit Knowledge and Presuppositions*. Chicago: University of Chicago Press.

Urquhart, C. (1997). Exploring Analyst-Client Communication: Using Grounded Theory Techniques to Investigate Interaction in Informal Requirements Gathering. In A. S. Lee, J. Liebenau, & J. I. DeGross (Eds.), *Information Systems and Qualitative Research* (pp. 149-181). London: Chapman and Hall.

Walsham, G. (1993). *Interpreting Information Systems in Organizations*. Chichester: John Wiley and Sons.

Walsham, G. (1995). Interpretive case studies in IS research: Nature and method. *European Journal of Information Systems, 4*(2), 74-81.

Walsham, G. (1997). Actor-network theory and IS research: Current status and future prospects. In A. S. Lee, J. Liebenau, & J. I. DeGross (Eds.), *Information Systems and Qualitative Research* (pp. 466-480). London: Chapman and Hall.

Walsham, G., & Sahay, S. (1999). GIS for District-Level Administration in India: Problems and Opportunities. *MIS Quarterly, 23*(1), 39-65.

Walsham, G., & Waema, T. (1994). Information Systems Strategy and Implementation: A Case Study of a Building Society. *ACM Transactions on Information Systems, 12*(2), 150-173.

Weber, M. (1930). *The Protestant ethic and the spirit of capitalism* (Parsons, Talcott, Trans.). London: Unwin University Books.

Weber, M. (1949). *The methodology of the social sciences*. Glencoe, Illinois: Free Press.

Winograd, T., & Flores, F. (1987). *Understanding Computers and Cognition: A New Foundation for Design*. New York: Addison-Wesley.

Wittgenstein, L. (1958). *Philosophical Investigations*. Oxford: Basil Blackwell.

Wynn, E. (1979). *Office Conversation as an Information Medium*. Unpublished PhD dissertation, University of California, Berkeley.

Wynn, E. (1991). Taking Practice Seriously. In J. Greenbaum & M. Kyng (Eds.), *Design at Work* . New Jersey: Lawrence Erlbaum.

Yin, R. K. (1994). *Case Study Research, Design and Methods*. (2nd ed.). Newbury Park: Sage Publications.

Ytterstad, P., Akselsen, S., Svendsen, G., & Watson, R. T. (1996). *Teledemocracy: Using Information Technology to Enhance Political Work* (1). MISQ Discovery. Available: http://www.misq.org/discovery/.

Zuboff, S. (1988). *In the Age of the Smart Machine*. New York: Basic Books.

Chapter X

Challenges to Qualitative Researchers in Information Systems

Allen S. Lee
Virginia Commonwealth University, USA

INTRODUCTION

A conventional "trends" chapter on qualitative research in information systems (IS) would review the state of the art (the methods and findings) of such research, laud its achievements, criticize its shortcomings, and then specify what it should do in the future to add to its achievements and rectify its shortcomings. However, I will write this chapter unconventionally instead, so that the reader will be able to gain a sense of my own engagement with issues in qualitative IS research. Furthermore, although the editor of this volume originally commissioned me to write a chapter on trends, the chapter has evolved as a critical commentary on qualitative IS research. The chapter's turn in this direction resulted from the editor's guidance to me about how to account for the comments of the anonymous reviewers of the initial draft.

I will proceed unconventionally in three ways. First, I will write in the first person, where my reason for doing so is not simply that the first-person style is gaining acceptance in academic prose, but also that my writing in the first person will, in itself, serve throughout this chapter as a continual reminder of my ontological belief about the nature of research and knowledge. I reject the naïve objectivist ontology that scientific knowledge exists or can be made to exist independently of knowing subjects. My belief is that all knowledge is a human creation and a human possession. In the same way that qualitative researchers often describe an organization, a custom, or a social practice as a socially constructed reality, I view scientific research as a socially constructed reality, where scientific knowledge is no less a human creation and possession than any other form of knowledge — a theme to which I will return continually in this chapter. For me to write solely in the third person or passive voice would be to pretend that I am invisible or have never existed. Such a pretense would be as silly as it is false. I will use the first person plural for referring to myself along with my colleagues in the community of qualitative IS researchers, and reserve the first person singular to express my own thoughts, value judgments, and experiences. However, I am writing this chapter for an audience that includes all IS researchers.

Second, I will apply our own qualitative IS research to ourselves as qualitative IS researchers. We are accustomed to being the observers, not the observed. Here, I will turn the tables on ourselves and force us to look at ourselves in the same way that we have investigated others. This will involve framing qualitative IS researchers (ourselves) as the users of a particular form of information technology (qualitative research methods) for managing a particular type of information (scientific theory and evidence). This framing will then position me to ask, "what lessons may I derive from our own past qualitative studies on information systems for insights about ourselves as qualitative IS researchers?"

Third, rather than write this chapter as an empirical account, I will present challenges to scholars who are members of the qualitative IS research community. I will make value judgments, offer negative criticisms, and propose new directions, all in a reasoned manner akin to that of an editorial or polemic.

I will begin by presenting my conceptions of what an information system is and what a systems approach involves, where I will also present the first of my challenges to the community of qualitative IS researchers. Next, I will proceed to take a systems approach in my examination of five published qualitative IS research articles. In each case, I will turn the tables on ourselves by applying the article's message about information systems to our own behavior as qualitative IS researchers. Each such application will yield another challenge. Last, I will offer my opinions on new future directions for qualitative IS research.

BASIC SYSTEMS CONCEPTS: A CHALLENGE TO PRACTICE WHAT WE PREACH

In my presentations and writings where I have addressed an audience of IS researchers, I have often included a review of what an information system is, including my own view of what a systems approach involves. Some members of my audience react as if they were seeing these concepts for the first time (they have said, "yes, I see that we have to position ourselves more in the intersection," referring to the oval in Figure 1, below), but there are others who gently chide me by asking why I would "lecture" them on something that they already know. To avoid any misunderstanding here, I will explain why I persist

Figure 1: A Way of Conceptualizing Information Systems

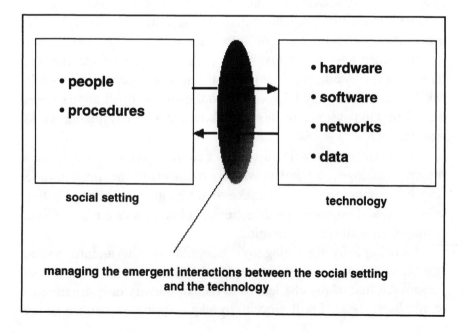

in the explicit attention I give to these basic concepts, especially when I am addressing an audience of IS researchers.

First, I am not convinced that all IS researchers realize what an information system is or what a systems approach involves. A situation manifesting this pertains to the emerging status of qualitative IS research in the overall IS research community. Qualitative IS research has succeeded in gaining acceptance in the formerly positivist-only bastion of North American IS research, but this has occurred only recently. Why did IS journals and conferences based in North America not accept qualitative research as part of the IS research mainstream all along? The reason I offer is that, for the most part, we qualitative IS researchers have not presented and positioned our research in a way that accounts for that fact that the research and publication process itself is a system—one with its own behavioral subsystem, technological subsystem, and behavioral-technological interactions, all of which are no less rich and consequential than what we commonly see in the information systems that we research in organizational settings. It is true that we qualitative IS researchers routinely take a systems approach in the task of researching people and information technologies in organizations, but this task is different from the one of presenting and positioning ourselves and our research to the larger IS research community. In failing to take a systems approach in the latter task, we have failed to practice what we preach. It is as if some interpretive IS researchers, knowing that their methodologies are correct (after all, the philosophy of science has disavowed positivism[1]), believe that the validity of their research will lead to its being published. In reality, the research and publication process has more than just a rational dimension; it is a system that involves not only technical rationality, but also behavioral practices.

I have encountered other situations also indicating a lack of appreciation, among IS researchers, of a systems perspective. At the May 1999 meeting of the Information Resources Management Association, I participated in a "meet the editors" session. As editor-in-chief of *MIS Quarterly*, I gave a presentation that included a diagram (Figure 1), taken from my first editorial statement in the *Quarterly* (Lee, 1999), depicting an information system as composed of a technological subsystem (the "technology"), a behavioral subsystem (the "social setting"), and the interactions that arise between them.

Despite this, one researcher offered a reaction (which he dissemi-nated to over 2,000 people on the IS World listserv) in which he characterized me as having said that *"MISQ* needs to function more like a social science journal and that, by design, *MISQ* does not involve itself with the IT side of the house"! (If he derived this statement from Figure 1, he could just as well have characterized me as having said that *"MISQ* does not involve itself with the behavioral side of the house.") This person's remark, in turn, reminded me of some dialogues I have had with certain other academic colleagues. In particular, I am thinking of, first, technocentric researchers who have no appreciation of the behav-ioral and, second, (to coin a new word) behavior-centric researchers (some, but not all, of whom come from outside the IS field) who have no appreciation of the technological. Members of the latter group believe that they can do IS research even though they have no particular expertise in technology, have never worked in the IS field, or have never taught an IS course. In dialogues that I have had with members of either group, I have noticed that their arguments have been consistent with the view that the behavioral and the technological are separable and therefore can be understood, researched, and managed independently of each other. My efforts to convey the importance of behavioral matters to technocentric researchers, the importance of technological matters to behavior-centric researchers, and (most difficult of all) the importance of behavioral-technological interactions to either group, have typically escaped these people. It is no wonder that some of the techno-centric and behavior-centric researchers, given their ignorance, routinely question the need for an IS discipline.

Researchers are not the only ones who have failed to appreciate a systems approach. A good example of this comes from my recent experience with a Fortune 500 company in the midwestern region of the United States. I was attending meetings of the company's senior executives involved in the implementation of enterprise resource planning (ERP) software. After I sat through several of their meetings, I came to realize that they were treating an information system as if it were a box containing hardware and software and, accordingly, what they felt they needed to do was to plug the box in, step back, and see it work. In years past, I had an MBA student who captured this technocentric thinking in a glib phrase; in a class discussion on a Harvard Business School case that I was teaching, she described the company in the case as having an "appliance mentality," where the people at this company treated their information system as if it were something you simply buy and plug in, like a refrigerator or washing machine, where no changes to the rest of the company's operations would be needed. And indeed, the "appliance mentality" explained what this Fortune 500 company was doing: the executives were treating every implementation problem with their new ERP system as if it were an appliance, like a washing machine, to be plugged in and turned on. If any changes would be required, the trick would be (as the executives said) to "configure" the ERP system to fit their company better. These executives gave no thought or consideration to changing their manufacturing or other business processes. In other words, the executives were not treating any of their implementation problems in a systemic, or systems oriented, way. For them, the second word in the phrase "information system" had no particular meaning, where the phrase was synonymous with "the computer." In this way, they exhibited a technology-only or technocentric view of an information system, where their lack of knowledge or acceptance of an information system *as a system* was counterproductive to their ERP implementation effort. All in all, these executives neither knew nor accepted a fundamental truth about information systems:

Figure 2: A Way of Conceptualizing IS Research *as a System*

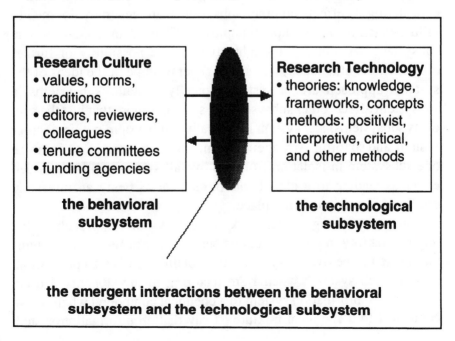

Research Culture
- values, norms, traditions
- editors, reviewers, colleagues
- tenure committees
- funding agencies

the behavioral subsystem

Research Technology
- theories: knowledge, frameworks, concepts
- methods: positivist, interpretive, critical, and other methods

the technological subsystem

the emergent interactions between the behavioral subsystem and the technological subsystem

namely, an information system *is a system*, whose interactive effects between the technological subsystem and behavioral subsystem require that people (here, the executives) understand and manage the two subsystems together, not separately.

Because of the persistence of people (both inside and outside of IS, and both inside and outside of academia) who simply "don't get it," I have made it a point to present my ideas explicitly on what an information system is and what a systems approach involves. (Conversely, it would be nothing short of a conceit for those of us IS researchers who do "get it" to suppose that everyone else already knows what we know.) To present my ideas, I have used the concept of an ecosystem to serve as an exemplar because it provides a graphic and concrete illustration of systems in general.

Almost like a laboratory experiment concocted to illuminate a point, an ecosystem well illustrates key features of systems in general. A deliberate change or intervention in one part of the system (such as the introduction of DDT) can ripple through the many subsystems constituting the overall (eco)system and eventually have all sorts of impacts, many of which are typically unforeseen and unpredictable and of which we typically become aware only after they manifest themselves. One subsystem can and does affect other subsystems, and any attempts to manage, intervene in, or optimize one subsystem, without considering its effects on other subsystems, are generally doomed to failure (such as an attempt to deploy DDT as an insecticide in one subsystem, without any thought given to the possibility of further effects on the other subsystems). In general, some of the ideas of the systems approach are: a system typically contains subsystems, a subsystem affects other subsystems, a system is more that just the sum of its subsystems, any system is typically a subsystem within a larger system, and so forth. Any attempt to manage, intervene in, or optimize one subsystem in isolation from the overall system can be regarded either as leading to a global suboptimum or as being an infeasible course of action in the first place.

In my thinking, a hallmark of any systems approach is the unpredictability in the interactions between the different subsystems that make up the overall system. In explaining my conception of an "information system" when I am lecturing to my students, I often compare an information system to a chemical compound where I also contrast it to a chemical mixture. I say that an information system consists of a behavioral subsystem and a technological subsystem, whose combined effects are not simply additive like those of different chemical elements that retain their respective properties when making up a mixture, but are more like those of different chemical elements that react to each other and *alter* each other's properties when forming a compound. For information systems in organizations, this means that the very same information technology can be a wild success in one company but a dismal failure in another — a fact still unacknowledged by much traditional, non-field based IS research.

A systems conception of information system is also instrumental for delineating how our IS field is different from the traditional behavioral disciplines (which focus on the contents of the behavioral-subsystem box) and the traditional technological disciplines (which focus on the contents of the technological-subsystem box). By my definition, our IS field does not deal with technology alone, or with behavior alone, or even with the simple concatenation of technology and behavior. Our IS field deals with the phenomena that emerge when the technological and the behavioral interact, much like different chemical elements reacting to one another when they form a compound. This is what makes our field different from the traditional behavioral disciplines and the traditional technological disciplines. I also rely on this argument to support my position that our overall IS field should stop referring to, and should instead declare independence from, any supposed "reference disciplines" (which include computer science, organizational theory, economics, and even operations research), none of which, in my view, have an adequate systems orientation.

On the one hand, the concept of a systems approach and the concept of an information system are not new or unique (but, I emphasize, they are not adequately disseminated or diffused either). On the other hand — and this is another theme of this chapter — these concepts pertain no less to people who are us (IS researchers, whether qualitative or quantitative), organizations that are our research institutions (including journals, conferences, and tenure committees), and information technologies that are our research methods. A challenge that I offer to the community of qualitative IS researchers is for us to practice what we preach by applying a systems perspective not only to the subject matters we are studying, but also to ourselves as researchers in how we present and position our qualitative research with the overall IS research community. Figure 2 attempts to capture the gist of this application.

The textual entries in Figure 2 are illustrative, not exhaustive. For example, there are more roles in the behavioral subsystem than just the "editors, reviewers, colleagues" that Figure 2 mentions. Another example is that the expertise or tacit knowledge that seasoned researchers bring to bear are no less a part of the technological subsystem than the explicitly mentioned "theories" and "methods."

In the argument that immediately follows, I will derive lessons and issue additional challenges based on my examination of five classic qualitative IS research articles. Bostrom and Heinen's "MIS Problems and Failures: A Socio-Technical Perspective" has become a classic through the endurance of its lessons (seven conditions leading to information systems failure) over the quarter century since its publication. Its lessons apply as much to today's distributed real-time ERP systems as they do to the mainframe-based batch transaction processing systems of concern to IS researchers in the 1970s. Markus' 1983 study, "Power, Politics, and MIS Implementation" is no less influential, having been cited by over 200 other articles since 1993 alone. Hirschheim and Klein's "Four Paradigms of Information Systems Development" is equally well known to qualitative IS researchers and was a watershed in the growing impact of philosophy of science on the discourse of North American IS research. Orlikowski's 1993 study, "CASE Tools as Organizational Change," is a landmark for having been the first qualitative article in *MIS Quarterly* to win its annual best paper award. Kumar, van Dissel, and Bielli's 1998 study, "The Merchant of Prato," also an *MIS Quarterly* best paper award winner, is significant for its empirical refutation of positivist transaction cost economics theory.

From Bostrom and Heinen's "MIS Problems and Failures: A Socio-Technical Perspective": A Challenge to Apply a Socio-Technical Perspective to Ourselves in the Research and Publication Process

In their *MIS Quarterly* article, Bostrom and Heinen pose seven conditions that lead to what they call "MIS problems and failures." Their Figure 1 (p. 21) contains the text reproduced on the left-hand side of the table that follows on the next page. In the corresponding text in the right-hand side, I am offering an application of Bostrom and Heinen's general ideas about people, information technologies, and organizations to the specific situation of people who are IS researchers, information technologies that are research methods, and organizations that are IS research institutions. I will employ the same rhetorical device with all five classic articles in my effort to derive insights that these past studies hold for ourselves as qualitative IS researchers.

I emphasize that, for each of the five articles, the application I provide will only be one possible application. Other applications, with additional insights, surely exist. The point of the rhetorical device is simply to show how the authors' insights about people, information technology, and organizational settings are also a source of insights about ourselves, our research methods, and our IS research community.

In the arena of IS practice, each one of the seven conditions (on the left-hand side) can contribute to the failure of an information technology in an organization. Likewise, in the arena of IS research and

Verbatim text from Bostrom and Heinen (p. 21) about people, information technology, and organizations.	An application of Bostrom and Heinen's text to IS researchers, IS research methods, and the IS research community.
Systems Designers' frames of reference [are] reflected in Seven Conditions:	Qualitative IS researchers' frames of reference [are] reflected in Seven Conditions:
1. "Implicit" theories, held by systems designers about organizations, their members, and how to change them.	1. "Implicit" theories, held by qualitative IS researchers about the overall IS research culture, IS research institutions, their members, and how to change them.
2. The concept of respon-sibility [for the succussful implementation of an information technology] held by systems designers.	2. The concept of responsibility [for the successful publishing of an IS research paper] held by qualitative IS researchers.
3. Limited conceptualizations of frameworks for organizational work sys-tems or user systems	3. Limited conceptualizations of frameworks for the publication system used by qualitative IS researchers

used by systems designers in the design process, i.e., non-systemic approach.	in the research process, i.e., non-systemic approach.
4. Limited view of the goal of an MIS implementation held by designers.	4. Limited view of the goal of the research and publication process held by qualitative IS researchers.
5. Failure of the systems designers to include relevant persons in the design referent group. Who is the user?	5. Failure of qualitative IS researchers to include relevant persons in the research process referent group. Who is the audience?
6. The rational/static view of the systems development process held by systems designers.	6. The rational/static view of the research and publication process held by IS researchers.
7. The limited set of change technologies available to systems designers who attempt to improve organizations.	7. The limited set of change technologies available to qualitative IS researchers who attempt to improve IS research.

publishing, each one of the seven conditions (as applied and presented in the right-hand side) can contribute to the failure of qualitative IS research to be accepted and disseminated. My experiences as an editor and reviewer have provided me opportunities to see how Bostrom and Heinen's seven conditions play out in the arena of IS research and publishing.

I have witnessed countless submissions of qualitative IS research papers to the refereeing process where the authors have presented their research with neither an elucidation of their particular qualitative method or an explanation of how their findings are valid. Background factors relevant to this situation are that qualitative IS research constitutes a minority in IS research overall and that qualitative approaches exhibit so much diversity that no single editor, reviewer, or other reader is or can be an expert in them all. These factors are well known to qualitative IS researchers. Therefore, I have found it astonishing that the authors of these submissions often make no attempt to communicate (i.e., explicitly describe the research method or show the soundness of the resulting conclusions) to the reviewers, the editor, or the eventual journal readers of their research. It is as if these researchers have not given careful thought to who their audience is or where their audience would be "coming from." When I am the editor for such a submission, I often ask the author to revise it before I send it to any reviewers, where I also say to the author: "Either you can be explicit in describing your qualitative research method and showing how your results satisfy one or another set of validity criteria, or you can leave it up to the imagination of the reviewers. I strongly recommend against doing the latter."

It is easy to apply Bostrom and Heinen's seven conditions in an interpretation of this situation. I see the authors of these qualitative IS research submissions as having a strictly rational view of the research and publication process — a process that, in actuality, has not only a rational dimension, but also social and political ones as well. The implicit theories that these authors hold about the overall IS research culture and IS research institutions include the beliefs that (1) editors, reviewers, and journal readers can be expected to know the research methods and validity criteria that they (the authors of the submissions) already know; and (2) the review process, being rational, will recognize or otherwise come to accept the validity of their research results. And if there is a reader who happens not to be familiar with the qualitative method used in the submitted paper, then these authors are behaving as if a mere citation to another study's explication of a particular qualitative method would induce the reader to feel sufficiently responsible to look up the citation, read it, understand its argument, and accept it! This strictly rational view indicates a limited conceptualization of the overall publication system. Authors who hold this limited conceptualization expect their research to be evaluated straightaway on its merits. They show no recognition of the additional social and political requirements of educating and courting reviewers and editors who, even if sympathetic to qualitative research, are often unfamiliar with it.

I sometimes offer the metaphor that qualitative and quantitative researchers proceed as if they were members of two different cultures who speak two different languages. (This image was more true ten years ago than today, but my own recent experiences as a qualitative author with quantitative reviewers and editors has assured me that there is still truth to my metaphor.) Yet, in the many qualitative IS research submissions that have traveled across my desk, I have observed that some qualitative IS researchers speak only their own language and behave only according to their own culture, despite the fact that their research will be encountering evaluators and other readers who speak a different language and behave according to a different culture. These authors fail to see the overall review and publication process as an occasion that, in effect, requires them to be bilingual and bicultural and that also poses not only the immediate goal of gaining an "accept" decision on their submission but also the larger goal of effecting change in the larger IS research community. Specifically, it is an occasion where they may proceed to use diplomacy in courting emissaries from a different culture, and even to forge an alliance where the two cultures (the quantitative and the qualitative) might even eventually reinforce and mutually benefit one another.

One might venture to say that, because Bostrom and Heinen's 1977 article is a classic, its seven conditions are so well known as to require no additional attention. On the other hand, the failure of many qualitative IS researchers to apply these concepts to themselves in the research and publication process is evidence these concepts are not as widely diffused as they should be. The challenge I present to qualitative IS researchers overall is for all of us not only to profess familiarity with basic socio-technical systems concepts, but also to internalize these concepts in our own beliefs about how to proceed in the research and publication process. I challenge my colleagues to become self-consciously bilingual and bicultural in interfacing with the larger, overall IS research community.

From Markus' "Power, Politics, and MIS Implementation": A Challenge to Recognize the Interactions Between the Behavioral and the Technological

In her 1983 *Communications of the ACM* article, Markus presented three theories of resistance to MIS. The abstract of her article appears in the left-hand side; an application of her ideas, in the right-hand area.

Verbatim text from Markus (p. 1201) about people, information technology, and organizations.	An application of Markus' text to IS researchers, IS research methods, and the IS research community.
Theories of resistance to management information systems (MIS) are important because they guide the implementation strategies and tactics chosen by implementers.	Theories of resistance to qualitative IS research are important because they guide the acceptance strategies and tactics chosen by qualitative IS researchers.
Three basic theories of the causes of resistance underlie many prescriptions and rules for MIS implementation. Simply stated, people resist MIS because of their own internal factors [Markus calls this the "people-determined theory"], because of poor system design [the "system-determined theory" or what I call the "technology-determined theory"], and because of the interaction of specific system design features with aspects of the organizational context of system use [the "interaction theory"].	Three basic theories of the causes of resistance underlie many prescriptions and rules for acceptance of qualitative IS research. Simply stated, people who occupy gatekeeper roles in journals, conferences, tenure committees, and other research institutions resist qualitative IS research because of their own internal factors [the "behavioral-subsystem determined theory"], because of poor research design and other technical flaws in qualitative research [the "technological-subsystem determined theory"], and because of the interaction of specific features in the "technological subsystem" of IS research with aspects of the "behavioral subsystem" of IS research [the interaction theory].

These theories differ in their basic assumptions about systems, organizations, and resistance; they also differ in predictions that can be derived from them and in their implications for the implementation process. These differences are described and the task of evaluating the theories on the bases of the differences is begun. Data from a case study are used to illustrate the theories and to demonstrate the superiority, for implementers, of the interaction theory.	These theories differ in their basic assumptions about research technology, research culture, and resistance; they also differ in predictions that can be derived from them and in their implications for the acceptance process of qualitative IS research. These differences are described and the task of evaluating the theories on the bases of the differences is begun. Data from the history of qualitative methods in IS research are used to illustrate the theories and to demonstrate the superiority, for qualitative IS researchers, of the interaction theory.

The "technology-determined" theory which Markus identifies (but, given the conclusion of her article, to which she does not subscribe) holds that the same information technology, if successful in the setting of one company, will also succeed in the setting of a different company. In contrast, the "interaction theory" which Markus identifies (and to which she does subscribe) holds that the same information technology, if successful in the setting of one company, does not necessarily succeed in the setting of a different company.

In transferring this insight from the process of implementing MIS to the process of publishing research, I identify the "technological-subsystem determined theory" as holding that the same qualitative research method (e.g., ethnography), if successful in the setting of one academic community (e.g., anthropology), will also be successful in the setting of a different academic community (e.g., the IS discipline). In contrast, the "interaction theory" would hold that the same qualitative research method, if successful in one academic community, does not necessarily succeed in another. Indeed, the history of qualitative methods in IS research refutes the technological-subsystem determined theory and supports the interaction theory. For years, qualitative methods that had been well accepted in other disciplines were dismissed in IS research.

As mentioned earlier, I have observed, in my roles as editor and reviewer, authors who attempt to justify one or another qualitative method or non-positivist approach simply by providing citations that refer the reader to studies published outside of the IS discipline. I regard these actions by these authors as evidence that their theory-in-use (Argyris and Schön, 1978) is the technological-subsystem determined theory, not the interaction theory. However, the technological-subsystem determined theory is wrong; a mere citation indicating the acceptance of a qualitative method in a non-IS field is insufficient to win acceptance of this method in the IS discipline. The challenge I offer to many of my colleagues who are qualitative IS researchers is for them to recognize the interactions between the behavioral and the technological, not only in the information systems that they observe in our research in organizations, but also in the research and publication process where the behavioral subsystem (including the political power and the different language of non-qualitative groups in the IS research world) has a role no less influential than any of the objective intellectual merits of a qualitative method itself. A qualitative method, depending on how it is positioned and justified, can interact with the behavioral subsystem (the overall, quantitative-dominated IS research community) so as to lead to the method's acceptance or rejection.

A brilliant research essay by Klein and Myers (1999) shows one way of carrying out this challenge: it provides a set of criteria, derived in part from published qualitative IS research articles, for assessing the validity of interpretive field studies. By grounding a qualitative method firmly in examples located in the IS discipline itself, it is building up credibility for the method in the IS research community. Also, instead of proceeding straightaway in doing qualitative research, the Klein and Myers essay is providing the necessary and prior justification which other qualitative researchers may subsequently cite when they go ahead and actually do interpretive field studies.

Five other recently published articles (Walsham and Sahay, 1999; Schultze, 2000; Trauth and Jessup, 2000; Gopal and Prasad, 2000; and Nelson, Nadkarni, and Narayanan, 2000) also show a way of carrying out this challenge: in addition to doing good qualitative research on IS topics, they also self-consciously formulate and apply evaluation criteria for judging the validity of their studies' conclusions, hence leaving nothing on this matter to the imagination of their reviewers, editors, or general readers. Furthermore, their criteria can serve future qualitative IS researchers, who may cite the earlier IS studies, present the criteria, and then explicitly justify the validity of their research, in ways that IS reviewers and readers would accept.

From Hirschheim and Klein's "Four Paradigms of Information Systems Development": A Challenge to Prescribe for Ourselves as We Prescribe for System Designers

Hirschheim and Klein's 1989 article in the *Communications of the ACM* is notable for its skillful application of philosophical concepts in IS research. As detailed in the left-hand side of the following table, they focused on the "paradigms" held by system developers. In the right-hand side, I take their thoughts one step further: I focus on the "paradigms" held by us IS researchers.

Verbatim text from Hirsch-heim and Klein (p. 1201) about people, information technology, and organizations.	An application of Hirschheim and Klein's text to IS researchers, IS research methods, and the IS research community.
The most fundamental set of assumptions adopted by a professional community that allows its members to share similar perceptions and engage in commonly shared practices is called a "paradigm." Typically, a paradigm consists of assumptions about knowledge and how to acquire it, and about the physical and social world. As ethnomethodological studies have shown… such assumptions are shared by all scientific and profes-sional communities. As developers must conduct	The most fundamental set of assumptions adopted by an IS research community that allows its members to share similar perceptions and engage in com-monly shared research practices is called a "paradigm." Typic-ally, a paradigm consists of assumptions about IS research knowledge and how to acquire it, and about the world of IS research institutions. As ethno-methodological studies have shown… such assumptions are shared by all scientific and professional communities. As IS researchers must conduct inquiry as part of their own

inquiry as part of systems design and have to intervene into the social world as part of systems implementation, it is natural to distinguish between two types of related assumptions: those associated with the way in which system developers acquire knowledge needed to design the system (epistemological assumptions), and those that relate to their view of the social and technical world (ontological assumptions).

…

The functionalist paradigm is concerned with providing explanations of the *status quo*, social order, social integration, consensus, need satisfaction, and rational choice. It seeks to explain how the individual elements of a social system interact to form an integrated whole. The social relativist paradigm seeks explanation within the realm of individual consciousness and subjectivity, and within the frame of reference of the social actor as opposed to the observer of the action. From such a perspective

research work and have to intervene into IS research institutions as part of doing research, it is natural to distinguish between two types of related assumptions: those associated with the way in which IS researchers acquire knowledge needed to do their research (epistemological assumptions), and those that relate to their view of their research culture and their research technology (ontological assumptions).

…

The functionalist paradigm is concerned with providing explanations of the *status quo* in how to do IS research; of social order in IS research institutions; of social integration across IS researchers; of consensus among IS researchers; of need satisfaction for IS researchers; and of rational choice in doing IS research. It seeks to explain how the individual elements of the IS researchers' social system interact to form an integrated whole. The social relativist paradigm seeks explanation within the realm of an individual IS researcher's consciousness and subjectivity, and within the frame of "social roles and institutions exist as an expression of the meanings which men attach to their world." The radical structuralist paradigm emphasizes the need to overthrow or transcend the limitations placed on existing social and organizational arrangements. It focuses primarily on the structure and analysis of economic power relationships. The neohumanist paradigm seeks radical change, emancipation, and potentiality, and stresses the role that different social and organizational forces play in understanding change. It focuses on all forms of barriers to emancipation – in particular, ideology (distorted communication), power, and psychological compulsions and social constraints – and seeks ways to overcome them.

reference of the individual IS researcher as opposed to the philosopher, sociologist, or historian as the observer of the practice of research. From such a perspective "social roles in the IS research culture and institutions in the community of IS researchers exist as an expression of the meanings which IS researchers attach to their research world." The radical structuralist paradigm emphasizes the need to overthrow or transcend the limitations placed on existing research-culture and research-technology arrangements. It focuses primarily on the structure and analysis of political power relationships in the IS research community. The neohumanist paradigm seeks radical change, emancipation, and potentiality, and stresses the role that different research-culture and research-technology forces play in understanding change. It focuses on all forms of barriers to emancipation of IS researchers – in particular, ideology (distorted communication), power, and psychological compulsions and social constraints – and seeks ways to overcome them.

Hirschheim and Klein make no claim that system developers have ever publicly espoused any of the four paradigms — the functionalist, the social relativist, the radical structuralist, or the neohumanist.

Rather, Hirschheim and Klein are portraying system developers as operating with one or another (to coin a new term) "paradigm-in-use," which I see as a special instance of a more general term that Arygris and Schön (1978) have already established, "theory-in-use." A person's theory-in-use is "the theory that actually governs his actions" (Argyris and Schön, p. 7), even if the person were to verbalize an altogether different "espoused theory" in response to a direct question about what explains his or her behavior. My classification of a paradigm-in-use as a specific case of a theory-in-use suggests that Argyris and Schön's ideas might help clarify the idea of paradigms held by IS researchers. To follow through on this suggestion, I must first briefly introduce the ideas of Argyris and Schön.

The people of interest to Argyris and Schön are professionals, such as physicians, accountants, architects, and managers. The topic of interest to Argyris and Schön is the effectiveness with which these professionals do their jobs. Argyris and Schön explain that, more often than not, a professional is not aware of her theory-in-use. (Note that a theory-in-use is not a scientific theory that academic researchers have formulated and tested, but refers more generally to whatever beliefs a person uses to make sense of his or her environment and to direct his or her actions in it.) A professional's theory-in-use typically directs her behavior subconsciously. For Argyris and Schön, the professional's being unaware of her theory-in-use can pose problems to her professional effectiveness, thereby creating the opportunity for benefit in her becoming aware of her theory-in-use and replacing it with a new theory. Of course, this is easier said than done. Often, a professional's theory-in-use is what Argyris and Schön characterize as "untestable" — that is, even when the professional encounters evidence in her daily experiences that would indicate her theory-in-use to be wrong, she retains the theory nonetheless. One factor contributing to this outcome is that the theory-in-use is subconscious, hence rendering errors in it undetectable. Another factor is that the theory itself is often so loose or inconsistent that it could successfully explain even opposite outcomes, hence making itself impervious to contradictory evidence; for instance, the racist theory-in-use, "blacks aren't as good as whites" could rationalize a black person's success ("he got the promotion because of affirmative action") as well as the same person's failure ("you see, more proof that blacks don't cut it").

When a professional is not conscious of his theory-in-use, he automatically responds to stimuli from his environment by using the same behaviors with which he has responded to the same stimuli previously. For routine tasks, the automatic and programmed nature of these responses is good. If a professional or any person had to reflect anew in order to formulate a response to every instantiation of the same routine task, his overall performance in the workplace would be poor. In general, the following process of accounting for stimuli from the environment is what Argyris and Schön call "single loop learning": if the person's path of travel is or will soon be "off course," he learns from feedback from the environment that he needs to correct his path so that he may proceed in the same direction as before. However, for tasks that are not routine but problematic, single loop learning can be dysfunctional: it would lead the person to continue in the same direction as before when, in fact, a different direction needs to be charted. In general, "double loop learning" refers to the corrective process in which the person becomes conscious of his theory-in-use, learns that it is directing him in the wrong direction, and devises a new theory to direct his actions. In this way, for situations where single loop learning is dysfunctional (i.e., the theory-in-use leads to ineffective and even self-defeating actions), double loop learning would be required. Of course, to the extent that double loop learning requires the person to become aware of his theory-in-use, it is not trivial to implement. As any clinical psychologist can attest, it is a challenge to get any person to become aware of his taken-for-granted beliefs, common sense assumptions, "scripts," or equivalently, what Argyris and Schön call a theory-in-use.

I believe that many (perhaps most) IS researchers, whether quantitative or qualitative, conduct their research without being aware of their paradigm. To paraphrase Argyris and Schön, an IS researcher's paradigm-in-use is the paradigm that actually governs her actions in the research and publication process. I interpret Hirschheim and Klein's four paradigms as manifesting themselves among IS researchers in the following ways.

First, some IS researchers (in particular, those in the dominant school of thought who have no reason to question the *status quo* in institutional arrangements in the IS research world) accept the existing research and publication process as "given" and act to maintain the *status quo* in the research and publication process. As authors, they themselves follow the formal and informal rules of the review process and, as reviewers with the power to recommend an acceptance or a rejection, they require authors to conform to the same rules. They require any new person entering this process (e.g., an assistant professor) to follow these rules, leading to this person's integration into the *status quo*. In fact, they also require a person who fails to conform to the rules to leave the research community through rejections of his journal submissions and eventually through the denial of tenure to him. They use not only the research and publication process, but also the entire process of doctoral education, to implement a consensus on the *status quo* in research and publication norms. I diagnose these researchers as subscribing to the first of the four paradigms; they are following a functionalist paradigm-in-use.

Next, there are some qualitative IS researchers whose actions follow from a focus on the consciousness and subjectivity of individual actors (editors, reviewers, and authors), where the meanings that these individual actors attach to their research world are manifested in the specific social roles and practices in the research and publication process. These researchers do not see the research and publication process in terms of the functioning of any large, overarching IS research institutions. Rather, they understand the research and publication process in terms of individuals. These researchers follow a social relativist paradigm-in-use.

Then, there are some qualitative IS researchers who see IS research institutions in terms of conflict and power, where the oppressive reign of quantitative researchers as gatekeepers in the publication process is to be challenged, undermined, and overthrown. I see these researchers as following a radical-structuralist paradigm-in-use.

Finally, there are those IS researchers whose paradigm-in-use is neohumanist. These researchers seek radical change and ways to overcome barriers rooted in distorted communication, power, psychological compulsions, and social constraints, so as to transcend unjust and unneeded limitations in the research and publication process.

In my view, a picture of the research and publication process consistent with the radical-structuralist paradigm would be quite accurate for the IS research community in the past. For journals based in North America, there was arguably an oppressive reign of quantitative researchers as gatekeepers in the IS publication process. Significantly, when the field of IS research began, research was simply known as "research"; there was no need for quantitative researchers to describe their work as "quantitative research" because there was no other widely accepted form of research. The term "research" meant "quantitative research"! In such a setting, it would be unsurprising for some qualitative researchers to attempt to challenge, undermine, or overthrow the gatekeepers. Indeed, among the qualitative and interpretive minority, the term "positivist" even acquired (and sometimes still carries) a pejorative connotation. However, I believe that, in conformance to some aspects of a portrayal consistent with the neohumanist paradigm, the IS research world has experienced some radical change and has witnessed the falling of "quantitative only" barriers (though, as a qualitative researcher myself, I believe that quantitative research still receives preferential treatment). For IS researchers not to become prisoners of paradigms-in-use that are no longer accurate, double loop learning (involving the discarding of an outmoded paradigm-in-use and its replacement with an accurate one) is required. In this setting, any attempt by qualitative researchers to challenge, undermine, and overthrow quantitative scholars who are, in fact, sympathetic and open to qualitative research would be counterproductive and dysfunctional.

Must an IS researcher be conscious of his paradigm-in-use? Just as (discussed above) there is not always a need for a professional to be conscious of her theory-in-use, there is not always a need for an IS researcher to be conscious of his paradigm-in-use. However, just as there are occasions where a professional's theory-in-use becomes inaccurate, ineffective, and dysfunctional, there are occasions where an IS researcher's paradigm-in-use becomes inaccurate, ineffective, and dysfunctional. I emphasize that, by "paradigm," I am not referring to any scientific theory that IS researchers might be developing or testing, but to their understanding of the IS research community in functionalist, social relativist, radical structuralist, or neohumanist terms.

Hirschheim and Klein demonstrated that paradigms-in-use direct the actions of system developers and that system development can stand to benefit by an appropriate change in a system developer's paradigm-in-use. The challenge is for us IS researchers not to hold ourselves above the system developers whom we often research, but to apply our insights about them to ourselves. The insight is that IS research can stand to benefit by an appropriate change in our paradigms-in-use. As for all cases of double loop learning, a nontrivial aspect of the challenge would be the task of becoming conscious of our paradigms-in-use in the first place.

From Orlikowski's "CASE Tools as Organizational Change: Investigating Incremental and Radical Changes in Systems Development": a Challenge to Design and Adopt a Strategy for Either Incremental or Radical Change in the Overall IS Research Community

In her article, Orlikowski examined the introduction of an information technology (in her study, it was CASE tools) in two companies. The abstract of her article appears on the left-hand side. In the right-hand side, I offer one possible way of transferring Orliskowski's insights to the situation of the introduction of a research technology (specially, qualitative methods) in two IS research communities (the North American and the European).

Verbatim text from Orlikowski (p. 309) about people, information technology, and organizations.	An application of Orlikowski's text to IS researchers, IS research methods, and the IS research community.
The findings of an empirical study into two organizations' experiences with the adoption and use of CASE tools over time are presented.	The findings of an empirical study into the North American IS research community's experiences and the European IS research community's experiences with the adoption and use of qualitative research tools over time are presented.
Using a grounded theory research approach, the study characterizes the organizations' experiences in terms of processes of incremental or radical organizational change. These findings are used to develop a theoretical framework for conceptualizing the organizational issues around the adoption and use of these tools — issues that have been largely	Using a grounded theory research approach, the study characterizes the two research communities' experiences in terms of processes of incremental or radical research-culture change. These findings are used to develop a theoretical framework for conceptualizing the sociological and political research-culture issues around the adoption and use of these qualitative research tools — issues that have been largely

missing from contemporary discussions of CASE tools.	missing from contemporary discussions of both qualitative and quantitative IS research methodologies.
The framework and findings suggest that, in order to account for the experiences and outcomes associated with CASE tools, researchers should consider the social context of systems develop-ment, the intentions and actions of key players, and the implementation process followed by the organization.	The framework and findings suggest that, in order to account for the experiences and outcomes associated with qualitative research tools, IS researchers should consider the social and political context of research development, the intentions and actions of key players, and the research and publication process followed by the overall IS research community.

Have qualitative IS researchers actually thought and acted in terms of "processes of incremental or radical research-culture change" in order to effect the acceptance and adoption of qualitative research? A technically sound CASE tool (or any other information technology) does not automatically receive acceptance in any company; to presume that it does would be technocentric. In the same way, methodologically sound qualitative IS research does not automatically receive acceptance in a predominantly quantitative IS research community; to presume otherwise would be no less technocentric than in the analogous situation for CASE tools.

The insights that Orlikowski's study suggests build nicely on those from Bostrom and Heinen, from Markus, and from Hirschheim and Klein. It can be helpful, but insufficient in itself, for us IS researchers to be aware that we need to apply to ourselves (and not only to the human subjects we observe) Bostrom and Heinen's socio-technical conditions, Markus' interaction theory, and Hirschheim and Klein's notions about paradigms. In addition, we qualitative IS researchers might also proceed to configure these ingredients into a coherent change strategy, whether incremental or radical, for transforming the IS research culture overall so as to make it more receptive to qualitative perspectives. This would require identifying key players in the IS research community, coming to an understanding of their intentions and actions, and designing one or another change strategy so that these key players and others in the future will be more receptive to qualitative research.

In looking back on the past, I cannot identify any concerted, broadly based effort among qualitative IS researchers to implement such change in the overall IS research community (although some hardworking qualitative IS researchers have done their best to raise consciousness to promote change on an individual basis). The continuation of such a trend on "automatic pilot" into the future would not be preferable to a new course of action where we consciously design and adopt a concerted, community-wide change strategy.

From Kumar, van Dissel, and Bielli's "The Merchant of Prato—Revisited: Toward a Third Rationality of Information Systems": a Challenge to Enter a Paradigm Based on Trust

In their study, Kumar et al. presented a new perspective for examining IS implementation. As the text on the left indicates, their empirical investigation involved an information technology, "SPRINTEL" and demonstrated the research benefits of taking a perspective focusing on trust, social capital, and collaborative relationships. In the right-hand side, I offer but one possible way of suggesting how Kumar et al.'s ideas can transfer to the situation of qualitative IS researchers,

Verbatim text from Kumar, van Dissel, and Bielli (p. 199) about people, information technology, and organizations.	An application of Kumar, van Dissel, and Bielli's text to IS researchers, IS research methods, and the IS research community.
The failure of SPRINTEL, an inter-organizational information system in Prato (Italy) raises a number of interesting questions with regard to the *technical-economic* and *socio-political* perspectives that currently	The failure of action research, a qualitative research method, to diffuse in the IS research community in North America raises a number of interesting questions with regard to the *technical-economic* and *socio-political* perspectives that currently dominate

dominate the information-systems/information technology literature. These questions underscore the importance of developing additional theoretical perspectives to help us better understand the role of information systems in organizations.

In this article we reflect upon these questions and their theoretical foundations in the context of a case study. The case study describes the implementation, usage and outcome of an inter-organizational information system. An analysis is made of the extent to which the *technical-economic* and *socio-political* perspectives are sufficient to explain the failure of this system. The outcome of the analysis shows that these two perspectives are insufficient to provide an explanation.

Based upon literatures from a variety of sources we develop a third, complementary, perspective. Like Kling's (1980) socio-political perspective, this perspective is also an interactionist perspective. However, instead of focusing on politics and

our perspectives on how and why qualitative research methods succeed or do not succeed in diffusing through the IS research community. These questions underscore the importance of developing additional theoretical perspectives to help us better understand the role of research methods in our research culture.

In this article we reflect upon these questions and their theoretical foundations in the context of a case study of action research among IS researchers in North America. The case study describes the situation in which relatively few North American IS researchers either understand or do action research. An analysis is made of the extent to which the *technical-economic* and *socio-political* perspectives are sufficient to explain the failure of this research method. The outcome of the analysis shows that these two perspectives are insufficient to provide an explanation.

Based upon literatures from a variety of sources we develop a third, complementary, perspective. Like Kling's (1980) socio-political perspective, this perspective is also an interactionist perspective. However, instead of focusing on politics and conflict as the primary interaction mode,

conflict as the primary interaction mode, it focuses on collaboration and cooperation as the key to understanding interaction processes. This perspective introduces a third rationality of information systems in which *trust, social capital,* and *collaborative relationships* become the key concepts for interpretation.	it focuses on collaboration and cooperation as the key to understanding interaction processes. This perspective introduces a third rationality of research development in which *trust, social capital,* and *collaborative relationships* become the key concepts for interpretation.

Just as a trust-based perspective in IS research can be innovative, a trust-based perspective among qualitative IS researchers regarding their position in the overall IS research community can be innovative. Hirchheim and Klein's four paradigms do not exclude a trust-based perspective, but do not explicitly mention it either.

Based on my own experience as a qualitative IS researcher, I would say that the time is right for qualitative IS researchers to entertain a trust-based perspective in dealing with the overall IS research community, including individuals who we feel to be "hard core" quantitative researchers. I recall a time when qualitative IS researchers, upon receiving unjust and uninformed rejections in the review process from quantitative reviewers and editors, typically reacted with an understanding of the situation that was adversarial, conflict-based, and Machiavellian. However, the IS research world has changed. Today, qualitative and quantitative IS researchers are even co-authoring studies (Trauth and Jessup, 2000), and quantitative IS researchers include qualitative articles as required reading for their IS doctoral students. Certainly, in situations where quantitative IS researchers are potential allies for qualitative IS researchers, a conflict-based paradigm-in-use would be counterproductive and self-defeating for qualitative IS researchers. Today, a trust-based paradigm for understanding the IS research community would be preferable and could promise to accelerate the acceptance of qualitative IS research.

CONCLUSION AND FUTURE DIRECTIONS

My examination of qualitative IS research, involving a reflective tour through five qualitative IS research articles, suggests that our theory (i.e., the body of qualitative IS research) can productively apply to ourselves, where the fit appears to be so good as to suggest pertinent questions about ourselves as qualitative IS researchers.

I have observed that qualitative IS researchers typically profess knowledge of what an information system is and what a systems approach involves. The overarching challenge is for all of us qualitative researchers to apply this knowledge not only when we observe users, managers, and information technologies in organizations, but also, when we ourselves interact with the larger IS research community. In this challenge, the ultimate information system is the one where the users and managers are ourselves, the information technology is our qualitative research methods, and the information is our scientific theory and evidence.

This challenge, in turn, invites others: to apply socio-technical principles (Bostrom and Heinen) to ourselves; to position our research to account for the interactions (Markus) between our research methods (an instance of the technological) and the research and publication process (an instance of the behavioral); to become conscious of our paradigms-in-use (Hirschheim and Klein) about how the IS research world works and to change to a new paradigm, as needed; to design and adopt a strategy for either incremental or radical change (Orlikowski) in the overall IS research community so as to increase its receptivity to qualitative research; and to entertain a paradigm based on trust (Kumar, van Dissel, and Bielli) in recognition of the emergence of quantitative IS researchers who are collaborating and allying with qualitative IS researchers.

The challenges can serve not only as advice, but also as a test of the validity of the qualitative IS research from which they derive. To the extent that these challenges embody and reflect past qualitative IS research itself, successful outcomes to the challenges would confirm the validity of the research while any less than successful outcomes would constitute reason for their revision.

REFERENCES

Argryis, C. and Schön, D.A.(1978). *Increasing Professional Effectiveness*, San Francisco: Jossey-Bass.

Bernstein, R.J. (1976). *The Restructuring of Social and Political Theory*, New York: Harcourt Brace Jovanovich.

Bostrom, R. P., and Heinen, J. S. (1977). "MIS Problems and Failures: A Socio-Technical Perspective." *MIS Quarterly*, 1(3),17-32.

Gopal, A. and Prasad, P. (2000). "Understanding GDSS in Symbolic Context: Shifting the Focus from Technology to Interaction," *MIS Quarterly*, 24(3), 509- 546.

Hirschheim, R. and Klein, H. K. (1989). "Four Paradigms of Informations Systems Development," *Communications of the ACM*, 32(10), 1199-1216.

Klein, H.K. and Myers, M.D. (1999). "A Set of Principles for Conducting and Evaluating Field Studies in Information Systems," *MIS Quarterly*, 23(1), 67-93.

Kumar, K., van Dissel H.G., Bielli, P. (1998)."The Merchant of Prato — Revisited: Toward a Third Rationality of Information Systems," *MIS Quarterly*, 22(2), 199-226.

Lee, A.S.(1999)."Inaugural Editor's Comments," *MIS Quarterly*, 23(1), v-xi.

Markus, M. L.(1983). "Power, Politics, and MIS Implementation," *Communications of the ACM*, 26(6), 430-444.

Nelson, K.M., Nadkarni, S., Narayanan, V.K., Ghods, M.(2000). "Understanding Software Operations Support Exerptise: A Revealed Causal Mapping Approach," *MIS Quarterly*, 24(3), 475-507.

Orlikowski, W. J.(1993). "CASE Tools as Organizational Change: Investigating Incremental and Radical Changes in Systems Development," *MIS Quarterly*, 17(3), 309-340.

Schön, D.A.(1983). *The Reflective Practitioner: How Professionals Think in Action*, New York: Basic Books.

Schultze, U.(2000). "A Confessional Account of an Ethnography About Knowledge Work," *MIS Quarterly*, 24(1), 3-41.

Trauth, E.M. and Jessup, L.M.(2000). Understanding Computer-Mediated Discussions: Positivist and Interpretive Analysis of Group Support System Use," *MIS Quarterly*,24(1), 43-79.

Walsham, G. and Sahay, S.(1999). "GIS for District-Level Administration in India: Problems and Opportunities," *MIS Quarterly*, 23(1), 39-65.

ENDNOTES

1 Schön (1983) quotes Bernstein (1976, p. 207): "There is not a single major thesis advanced by either nineteenth-century Positivists or the Vienna Circle that has not been devastatingly criticized when measured by the Positivists' own standards for philosophical argument. The original formulations of the analytic-synthetic dichotomy and the verifiability criterion of meaning have been abandoned. It has been effectively shown that the Positivists' understanding of the natural sciences and the formal disciplines is grossly oversimplified. Whatever one's final judgment about the current disputes in the post-empiricist philosophy and history of science … there is rational agreement about the inadequacy of the original Positivist understanding of science, knowledge and meaning."

Chapter XI

Choosing Qualitative Methods in IS Research: Lessons Learned

Eileen M. Trauth
Northeastern University, USA

INTRODUCTION

This book is about the use of qualitative methods in the conduct of information systems research. As the title suggests, it is concerned both with *trends in the choice* of qualitative methods and with *issues with the use* of these methods. The issues have been addressed on two levels. The section on individual issues considers specific issues encountered by individual researchers in the conduct of particular research projects. The section on issues for the profession considers issues that the IS profession is currently confronting and those it will have to address in the future.

The subject matter of this book has considered the use of qualitative methods in information systems research in the following way. This book began by considering trends in the choice of qualitative methods: why and how people have chosen qualitative methods for the conduct of information systems research (Trauth, Wynn). It then moved on to consider individual issues. These issues are "individual" in two ways. They relate to an individual research project. They also relate to an individual qualitative research method. As the book moved through consideration of individual issues, the reader could note how qualitative methods have been used with a range of epistemologies. Beginning with qualitative methods employed in multi-method, action research (Mumford), the chapters moved into interpretive (Schultze, Urquhart) and critical (Cecez-Kecmanovic) research before coming full circle to consider another use of qualitative methods in multi-method research (Sawyer). The consideration of issues for the profession includes both philosophical and practical issues that the IS profession needs to address in order for qualitative methods to flourish. These include the risks involved in choosing qualitative methods (Baskerville), the need for a cumulative tradition of qualitative IS research (Klein and Myers), and finally, a range of future challenges for qualitative researchers in the IS field (Lee).

Building on the framework developed by Klein and Myers, this chapter presents lessons about the choice of qualitative methods for IS research that are found in this book. These lessons are presented according to their philosophical, social, methodological and political dimensions. Using this framework I provide both a summary of the key themes presented in this book and a taxonomy of lessons for those who would like to choose qualitative methods for their information systems research.

LESSONS LEARNED

Philosophical Dimension

Lesson 1: Separating Epistemology from Method

As these chapters have illustrated, it is important to distinguish the philosophical underpinnings from the methods that are employed to enact them. A given epistemology may employ a variety of methods just as a particular method may be employed in research that reflects different epistemologies. It is also necessary to distinguish the quantitative-qualitative dichotomy from the positivist-non-positivist debate. The use of qualitative methods does not necessarily imply interpretive research. For example, a qualitative method such as interviewing may be used for positivist, interpretive or critical research. Similarly, one can develop hypotheses while employing qualitative methods and use them for testing theory.

Nevertheless, there are important areas of interaction between epistemological theory and methodological choices. For example, Sawyer found that developing a multi-method research approach also required him to confront epistemological choices. In chapter 1, I noted that the epistemological orthodoxy of a particular context may determine which methods are deemed acceptable to use. Further, as Urquhart points out, the resolution of practical problems, in turn, leads to philosophical considerations. As she learned in her exploration of grounded theory literature, choosing an approach requires knowledge of the full body of work in order to understand the different philosophical perspectives. In order to employ grounded theory, she needed to know the debate and the implications for practical implementation in method. Understanding the philosophical perspectives enabled her to cope with the contradictory advice that resulted from these philosophical differences.

Social Dimension

Lesson 2: Using Social Theories

Whereas the first lesson makes the connection between *epistemological theory* and research method, this lesson is about the connection between *social theory* and method. Cecez-Kecmanovic's chapter demonstrated how linking social theory and method is an ongoing process. She found that the choice of methods and decisions about the conduct of the research could not be separated from the theory informing the research. The relationships among the social theory, research questions and empirical methods surfaced as a key issue for her research strategy. In the end, the methods that were chosen flowed from the social theory (Habermas) and the epistemology (critical). For Sawyer the connection is that a priori social theory helped him structure his data-collection process. Klein and Myers remind us that social theory needs to be confronted with the results from empirical studies.

Lesson 3: Embracing a Socio-Technical Perspective

To recognize the need for qualitative approaches to IS research is to recognize that an information system is a socio-technical system. This means that the information system designer considers the design of organization structures in addition to the design of information technology. To accept the inclusion of organizational context in information system design means that the designer must have an understanding of the context or situation before she or he can begin to solve technical problems.

This recognition carries with it implications for IS research topics and IS research methods. According to Baskerville and Lee this approach reflects the belief that human organizations interacting with IT can only be understood as whole entities. If the technological and the behavioral are not separable, by implication they should not be researched separately from each other. As Mumford points out, this recognition, in turn, introduces a political element into a heretofore technical domain.

Once the journey down the road of context is undertaken, another research implication is the recognition that information systems research is concerned with human context at varying levels of analysis. These levels—societal, cultural, industry—must be explored and understood in order to understand design issues related to technological societies. This theme was expressed by Mumford and echoed in Walsham's (2000) argument for greater breadth in the context dimension of IS research.

Several authors make a link between embracing a socio-technical perspective on information systems and adopting an interpretive stance in IS research. If much about IS involves change resulting from use of technology, then according to Schultze, interpretive methods enable study of the "sense making" that accompanies such change. Work practices need to be better understood for successful IT implementation, and through interpretive methods we can get at the messy reality of these practices. Since technology is socially constituted, she concludes, qualitative methods help us grasp the influence of social environment on technology.

Finally, as Klein and Myers point out, by choosing this research approach, totally new types of information regarding information systems development and use are possible. For example, my interpretive analysis of group support system use produced a distinction among three different types of information—cognitive, affective and behavioral—something that was not possible with the positivist analysis of the same data (Trauth and Jessup, 2000).

Lesson 4: Acknowledging the Social
Construction of Research and Knowledge

Just as information systems are situated within a social context, so too is knowledge situated within a social context. Lee positions his challenges for the qualitative IS research community against the backdrop of a socially constructed reality: scientific knowledge is not divorced from knowing subjects. Thus, social systems—organizations, customs and social practices— are viewed as a socially constructed reality. The link to qualitative methods, then, is clear to Schultze and Wynn: information creation is a social process and qualitative methods are better suited to getting at social processes, in all their complexity, than are quantitative methods.

Recognizing the social construction of knowledge leads naturally to the appeal of participant observation, which is employed in a range of qualitative methods including action research, ethnography and case study. But as Mumford points out in her discussion of the reality of conducting action research, a host of new challenges confront the researcher. Characteristics of the research setting such as gender attitudes might affect who can do what research and how. The researcher may also need to address ethical issues that surround the research. For example, in my study of Ireland's information economy, I encountered the ethical dilemma of how much of my own thoughts and feelings I should disclose to my research "subjects" as I got to know them better (Trauth, 1997).

Lesson 5: Recognizing the Primacy of the Researcher

With qualitative research the role of the researcher is at the same time different from and more prominent than what is the case for quantitative research. The researcher is not a detached observer; rather that person becomes, in all his or her human variety, the research instrument. Several of the authors (Baskerville, Cecez-Kecmanovic, Mumford, Schultze and Urquhart) noted that with the various types of qualitative research, the characteristics of the researcher move to the foreground. Because the observer's values and a priori knowledge help to shape the interpretation, the person(ality) of the researcher becomes a significant factor in the findings that result. What the researcher brings to the research setting, characteristics such as her or his assumptions, age, and gender, are all part of the research process. Whereas positivist researchers would consider the interjection of the researcher into the research setting as a disturbance to be overcome, interpretive and critical research consider the conscious recognition of the researcher to be essential to the process.

Sawyer reminds us that recognizing the primacy of the researcher brings with it the need to recognize how his or her total presence affects the "subjects." This, in turn, introduces new skills that this "research instrument" must possess. Mumford points to the multiple competencies that are needed: the ability to learn from experiences, interpersonal skills, motivational skills and creativity. Further, qualitative researchers need the ability to establish and maintain good personal relationships with subjects because negotiating access to "the data" is an ongoing process. What this means is that qualitative researchers need social skills in conducting this research, to put the interviewee at ease. Mumford emphasizes the need to establish credibility with subjects. In her work she needed to gain the acceptance of categories of respondents, which ranged from trade unions to dockers to miners. To her, a significant part of gaining credibility with respondents was achieved by maintaining confidentiality.

A final aspect of the researcher as "research instrument" that surfaced in these chapters is that this is not a completely solitary activity. There is the need for access to a supportive thought community to help the researcher process and write up the results of the study. Schultze describes her weekly meetings with her dissertation advisor who helped her think through her emerging interpretations. She also discusses the contribution of a writing seminar in which she was able to share and receive feedback on her interpretations. Urquhart talks about the need for coding seminars when engaging in grounded theory research. These forums provide mutual support to people engaged in open coding. An aspect of the solitary "research instrument" that arose in two chapters is the issue of the qualitative researcher working with others. While most of the chapters considered the researcher as an individual working more or less alone, Sawyer (and Mumford to a lesser extent) discuss issues for a team engaged in qualitative methods. In particular, Sawyer asks questions about a *team* as the research instrument. If a single researcher's values are important in the collection and processing of field data, how does a team behave when values and other such personal characteristics are different across members of the research team?

Methodological Dimension

Cecez-Kecmanovic needed to define methodology broadly as an overall strategy for conceptualizing and conducting research. For her, the issues went beyond consideration of particular empirical methods. Thus, while some of the methodological lessons considered here are about understanding and choosing a particular method, others are about the challenges of collecting the data and interpreting the results.

Lesson 6: Using Various Qualitative Methods

As Wynn points out, qualitative research in the information systems field represents the importation of several different methodological traditions. What they all share in common, however, is the focus on language and rich description as the mechanisms for analyzing the behavior of information systems and technology within their context of use. The chapters in this book illustrate the array of qualitative techniques that are available for IS research. This should enable the researcher to keep at the forefront the criteria for choosing a method. The deeper understanding of each method provided by the chapters in this book enables the reader/researcher to consider which of the various alternatives fits the research problem best.

Several authors cite the need that researchers – in particular Ph.D. students – have for more examples of the various qualitative methods. And as Klein and Myers point out in the introduction to their classification scheme, we are approaching a critical mass of such work. This and similar books (e.g., Lee and Leibenau, 1997; Mumford, et al., 1985; Nissen, et al., 1991), along with special issues of journals (e.g., Lee and Markus, 1999-2000; Myers and Walsham, 1998), provide a substantial number of methodological examples for the qualitative researcher to emulate. This body of work allows for a closer examination of particular areas of research in order to discern which ones are well developed and which are not. It also allows for division of labor among qualitative scholars and enables greater refinement and focusing of qualitative work. Finally, these examples provide not only procedural advice, but as Lee observes, they also show how to address the validity of findings from qualitative research.

Participant observation is the underlying approach in the various qualitative methods that are discussed in this book. The choice of how to apply participant observation techniques derives from decisions about active versus passive intervention in the context under study.

Action research represents the decision to intervene and comment on the change process in social systems. As Baskerville and Mumford point out in their chapters, it is a collaboration with subjects in the joint solution of real problems. It merges research and practice to engage in practical problem solving and to produce relevant research findings. The benefit of such a collaboration is that it enhances the competencies of the respective actors: the researcher and the client.

A contrasting use of participant observation, as exemplified in *ethnographic research*, assigns to a researcher the passive role of observing and recording changes. Schultze considers the opportune use of this interpretive method to be research settings that demand an understanding of context and all its complexity. For her, the choice of ethnographic methods arises when problems are unclear and complex or when the phenomenon is embedded in the social system, is poorly understood or is unknown. This interpretive method is suitable for studying work practices as "lived work" in everyday, mundane detail and for studying the structuring influence of social action on the use of information technology.

An alternative to the interpretive use of qualitative methods is provided by *critical research*. Cecez-Kecmanovic explores the nuances of this emerging alternative to positivist IS research. Using this approach the researcher helps practitioners and users understand the social consequences of information systems and helps them to consider alternatives. In doing so it exposes inherent conflicts, agendas of privileged groups and contradictions, which account for the influences exerted by context. Insofar as critical research speaks for users, this method gives voice to others than management.

Whether the researcher is choosing an active or a passive stance in the participant observation, qualitative methods can become part of a *multi-method research* approach, as Sawyer illustrates in his chapter. In his view, the use of multiple methods allows for a more robust and richer picture of events than a single method might allow. It enables the researchers to see the same event from different perspectives and facilitates multiple levels of analysis.

Lesson 7: Being Engaged in the Research Setting

Along some dimensions the researcher's engagement in the research setting is the same whether one is an active or a passive participant observer. First, with all types of qualitative research, the researcher has an intimate, long-term relationship with the research situation. Therefore, as Schultze discovered, one has to possess both the willingness and the ability to become physically, intellectually and emotionally immersed in the situation under study. As she explains, this often means giving up control. I also found in my research on Ireland's information economy, that I was not in control over when and where or what data I would be collecting (Trauth, 1997). The lesson is that the participant observer is vulnerable. Second, maintaining access to the data is both intellectual and emotional work. Conducting in-depth research exacts a time, energy and personal toll on the researcher. Time and energy are required both to collect the data and then to write the results. Mumford, Sawyer and Schultze all comment on the intense energy involved in fieldwork; my own experiences echo this sentiment. Finally, this intimate connection to the research context brings with it ethical responsibilities. Mumford emphasizes the ethical issue of not betraying confidences of the respondents in publishing research results. As I became friendly with some of the respondents in my Irish study, I encountered the ethical dilemma of when and whether to be always "at work" and recording the insights I was gleaning (Trauth, 1997).

Along other dimensions the researcher's engagement with issues will change, depending upon the nature of one's participation in the research setting. For some, there is a "switching cost." Sawyer considers the challenge of being immersed in more than one data set at a time and coping with the personal "switching cost" associated with multi-method research. Schultze also encountered a "switching cost" – that of maintaining both proximity to the data (needed for the participant observation) and distance from it (needed for reflection and analysis) when conducting reflexive ethnographic research. With ethnographic research one is always physically engaged with the research setting. But she grapples with determining when she must also be engaged emotionally, psychologically and intellectually. That is, she must determine when she *is* and when she *is not* collecting data.

For others, engagement issues will flow from the decision about whether to actively engage with the topic in situ or to passively observe the effects of IT in the workplace setting. Mumford notes that with action research the researcher is vulnerable to political and practical problems. She describes the challenge of negotiating volatile political situations, varying hidden agendas and the potential for management censorship of results that are to be published.

Regardless of the activity or passivity involved in the participant observation, the authors of this book all share practical issues they encountered and how they addressed them. For example, Mumford tells us about her technique of using perfume to signal her presence to the all-male coal mining community she was about to enter. I once relied upon interview venue (a pub) and dress (blue jeans) in order to establish more rapport with one of my respondents in an interview setting.[1]

Lesson 8: Working with the Data

The choice of a qualitative method and the ensuing type of engagement in the field leads naturally to lessons about processing the data. Whereas quantitative data is suitable for computerized analysis, qualitative data requires that the researcher do the analysis. Only one contributing author (Sawyer) mentioned the use of analysis software and he is moving away from it because "... the hard work is developing the structures and themes not keeping track of them."

Because the researcher is doing the analysis, qualitative research demands that she or he develop a considerable level of familiarity with the data. There are several implications that flow from the importance of knowing one's data. Sawyer notes the additional (mental and temporal) burden in multi-method research of learning multiple dissimilar data sets and working through the interim theories that arise from the data. But even with single method research, there is an issue of coping with the volume of data that exists when one is working with words instead of numbers. It is more difficult to organize and present the data in a way that allows one to comprehend the data set. In my own case, the findings of my interpretive investigation of Ireland's information economy required a 458-page book!

Conducting qualitative analysis requires considerable mental flexibility. The willingness and ability to engage with the data are necessary conditions for the deep and consuming reflection that is needed to carry out the analysis. Urquhart speaks of "living with the data for a long time" while Sawyer speaks of long walks as a technique for facilitating deeply reflective analysis. Schultze identifies the ability and tenacity to carry on amidst ambiguity as another necessary condition for qualitative analysis.

The challenge of coding the data for analysis is not only about the volume of data involved. It is also about whether to take a bottom-up or a top-down approach to the coding. Whereas a top-down approach using existing categories might make the process easier, the research circumstances sometime call for taking an inductive approach to empirical data. While many different types of coding exist, one type of coding which employs the inductive approach is *grounded theory*. Urquhart's chapter focused explicitly on the analytic technique of grounded theory; it is implied in other chapters. Grounded theory is applicable when there is little theory available and when an exploratory and inductive approach to the data is appropriate. The coding involved with grounded theory is done at an analytic level, when the research and the data are driving the coding paradigm that is chosen. As she illustrates, there is considerable rigor involved in this method of qualitative data analysis.

A derivative issue related to grounded theory is that of dealing with the existing literature on the topic. While some incorrectly infer that an inductive mode is an injunction against reading literature when engaged in grounded theory, Urquhart explains that the researcher is simply relating to the literature differently. The literature is read after the substantive theory has been developed.

The role of "writing up the results" when conducting qualitative research is also different. Schultze explains that writing is both analysis and presentation of results. Thus, in many respects the writing *is* the analysis. Therefore, writing style has implications for both what is said and how data is analyzed. In my experience of conducting qualitative analysis, writing has served as a vehicle for facilitating the thinking process. The discipline of writing down one's thoughts forces the researcher to think through the topic. Urquhart mentions the role of theoretical memos in her research. These memos serve as early versions of the final written product.

Lesson 9: Maintaining an Adaptive Stance

A final lesson related to the methodological dimension of qualitative research is the adaptive stance that is implied in this research approach. Qualitative methods are particularly suitable when we do not understand the dimensions enough to express them as quantifiable variables. Whereas quantitative research fails us when we attempt to measure the unknown, the uncertain or the serendipitous, the adaptive feature of qualitative research enables it to rise to the challenge.

This adaptive feature is present in every qualitative method whether it is action research, ethnography or critical research. Mumford expresses it as learning by doing and getting better with experience. Schultze speaks of gaining unplanned insights that emerge as events occur in their natural setting. She describes data collection as constant data analysis, a research approach that is also loose and messy at times. As Urquhart notes, the ability to develop meanings in grounded theory is *only available* with qualitative methods because they afford the opportunity to consider how meanings evolve and change. The meanings and theory emerge from the data through constant comparison: through labeling and analysis. With grounded theory even the research questions are flexible. Urquhart shows how one doesn't start out with the research questions; instead, they result from the analysis.

Two keys to the adaptiveness inherent in qualitative analysis can be found in ongoing analysis and in contradictions. Sawyer talks of the need for ongoing analysis and the development of interim analyses. I have found contradictions to be a rich source of insight regarding assessment and possible redirection. Unlike positivist research where anomalies are something to be avoided, both ethnographic and hermeneutic research employ anomalies as central analytic devices. In our study of group support systems we focused on breakdown analysis (Agar, 1986) to show how our insights and interpretations of the data emerged (Trauth and Jessup, 2000).

One way in which the adaptive feature of qualitative research is demonstrated is through the reflexivity that this method allows. Schultze relied heavily on reflexivity in her analysis of the informing behavior of knowledge workers. The lesson she learned was that in order to successfully employ this approach to qualitative research one needs both a willingness and the ability to write self-reflexive field notes.

Finally, Cecez-Kecmanovic points to reflexivity as part of the reflective thinking that is inherent in qualitative analysis. She warns that while reflective practices are good in any research, they are critical in some. In both interpretive and critical research, "the truth" is not something "out there" waiting to be discovered. Rather, through reflexive and dialectical processes, researchers become more explicit about their assumptions and what they bring to the research process than would be the case in positivist work. This, in turn, helps the researchers to understand their own engagement with their subjects. It also enables researchers to make connections to past experiences.

Finally, it enables a team of researchers to develop mutual understanding and provides a basis for exploring differing interpretations.

An adaptive stance is needed not only in the execution of a particular qualitative method, it is also needed in one's approach to qualitative methodology in general. Qualitative research methods are changing as the field of information systems discovers and imports new methods into it. As more people use them, we can build upon this knowledge. Cecez-Kecmanovic notes that methods evolve as our understanding of a particular topic emerges. Further, by identifying methodological limits we point to areas requiring further research into methodology.

Political Dimension

The political dimension of qualitative research is played out in both the corporate politics of the research site and in the academic politics of the researcher's career. According to Lee, paradigms-in-use govern one's actions in the research and publication process, and must be altered when they are no longer valid. In his view, the image of the beleaguered qualitative researcher is no longer accurate. Nevertheless, he argues, in order to be successful one has to develop an emerging and accurate understanding of the IS research community.

Lesson 10: Coping with the Risks Involved
with Using Qualitative Methods

Part of this political understanding is an understanding of the risks involved. As Wynn explains, the issue of legitimacy is about the transportation of qualitative methods that are totally acceptable in one discipline (such as anthropology) into another field that has been less receptive (such as information systems). This is Baskerville's risk of encountering challenges to the underlying data and analytical techniques being used. The philosophical and methodological risks are the possible rejection of results because the researcher did not engage in traditional science: deductive logic, problem reduction and detached objectivity. The "soft data" that results lacks representative samples and statistical generalizations. To counter this risk Lee suggests the need to formulate, articulate and apply one's own evaluation criteria. Indeed, one of the contributions of the Special Issue of *MIS Quarterly* on Intensive Research (1999-2000) was to develop and present evaluative criteria for interpretive research.

While Baskerville's consideration of risks is associated with action research, many of his recommendations can be extended to other areas of qualitative research as well. Failure to contribute substantially to theory results from acquiring new knowledge without developing a useful definition of what was learned. Outcomes failure is the failure to achieve stated purposes or level of quality. Both Baskerville and Mumford point out that for action research this means making sure that the solution addresses the client's problem as well as adding to IS theory. The highest risk with qualitative as with quantitative research is that neither practice nor science is served by the results.

There are several causes of outcomes failure. One occurs when the researchers disengage from the research too early. Another results from a lack of collaboration between subjects and researchers. In the case of action research this might be an imbalance between the creation of general knowledge and the fixing of practical problems. In the case of ethnography it might mean a disconnect between the ethnographer and the respondents. A third risk is detachment, the researcher becoming detached from either the scientific or the practical goals. The final risk is that of abstraction. Researchers who are trained to value the contemplative act must behave differently and be engaged in the field.

Lesson 11: Surviving Academic Politics

Qualitative research is not conducted in a vacuum. As Chapter 1 points out and Schultze and Urquhart illustrate, the choice of method is influenced by a range of personal and political factors that range from family responsibilities to teaching schedules to tenure clocks to the acceptability of such methods at one's university. From the experiences of the contributing authors, it is clear that different parts of the world have widely different attitudes about "acceptable" IS research methods. For example, Baskerville points out that in Europe action research has been closely linked with (and is viewed as acceptable for) system development research. He cites Checkland's (1981) soft systems methodology as a case in point.

Much about the academic politics surrounding qualitative research involves the politics of publishing. Baskerville explains how the academic culture in IS research can lead to methodological risk avoidance. In the beginning of one's career the desire for a low risk dissertation might keep one within the quantitative mainstream. Then, as a junior faculty one can become trapped within the expedient route toward tenure: writing journal articles from that quantitative thesis. Finally, post-tenure pressure to maintain one's position in the field puts pressure to remain in the methodological mainstream. Thus, the "publish or perish" mandate of an academic career can dissuade one from ever choosing qualitative research.

Others pick up on the pressures, especially on Ph.D. students, that are associated with the choice of qualitative methods. The deep reflection — the "long walks" of Schultze and Sawyer – needed for qualitative analysis of data requires having large mental and temporal spaces for reflection. But this may run counter to an increasingly sped up world and educational plan. Further, analytic and presentation techniques such as writing self-reflexively have not normally been part of a Ph.D. curriculum. Instead, Urquhart points out that one is generally directed to "focus" in one's dissertation, and, therefore, reject a grounded approach.

Among the challenges associated with publishing qualitative research are such pragmatic issues as the constraint of page length. When the data being analyzed are words, not numbers, more space is generally needed to explain the methodology, results and criteria for evaluating those results. Books are often the appropriate vehicle for presenting the results of interpretive studies, for example. Yet, the IS establishment is not always receptive to books as a substitute for journal articles. Further, within the realm of papers, page lengths that assume a positivist research approach can further serve as an inhibitor to qualitative writing.[2]

Responses to this situation are suggested in the chapters by Lee and by Klein and Myers. Lee suggests that qualitative researchers adopt the same socio-technical stance both in the conduct of the research and in the publication process. That is, the researcher needs to know who the audience is and where they're coming from. He argues that a solely rational view of the publication process that doesn't incorporate the social and political views impedes publication of qualitative research. Political power and language influence the publishing of qualitative work and must, therefore, be addressed. So, for example, it becomes incumbent upon the qualitative researchers to speak to those who speak another language: the quantitative majority. In essence, qualitative researchers should teach editors and reviewers about qualitative methods.

Klein and Myers' response is to recommend the development of a cumulative tradition in the literature. They argue that typologies are needed in order to build a cumulative tradition. These typologies provide guidance about empirical, interpretive research design and about writing up the results. They also help the research with respect to focusing the work and characterizing one's contribution.

CONCLUSION

This chapter concludes the section on issues for the profession as well as this book. It offers a set of lessons about the use of qualitative methods, which are culled from the chapters in this book. Whether the learner is a graduate student in a structured course or a scholar learning independently, the lessons about the choice of qualitative methods are intended to help future qualitative researchers anticipate some of the issues they might confront. This will enable them to better cope with the challenges of choosing qualitative methods for information systems research.

REFERENCES

Agar, M.H. (1986). *Speaking of Ethnography*, Sage Publications, Newbury Park, CA.

Checkland, P. (1981). *Systems Thinking, Systems Practice.* New York: John Wiley and Sons.

Lee, A.S., Liebenau, J. and DeGross, J.I. (Eds.). (1997). *Information Systems and Qualitative Research,* London: Chapman & Hall.

Lee, A.S. and Markus, M.L. (Eds.) 1999-2000. *MIS Quarterly* Special Issue on Intensive Research, 23(1), 24(1) and 24(3).

Mumford, E., Hirschheim, R.A., Fitzgerald, G. and WoodHarper, T. (Eds.)(1985). *Research Methods in Information Systems*, Amsterdam: NorthHolland.

Myers, M. and Walsham, G. (Eds.) (1998). *Journal of Information Technology Special Issue on Interpretive Research*, 13(4), 235-246.

Nissen, H.-E., Klein, H.K. and Hirschheim R. (Eds.) (1991). *Information Systems Research: Contemporary Approaches and Emergent Traditions*. Amsterdam: North-Holland.

Trauth, E.M. (1997). Achieving the Research Goal with Qualitative Methods: Lessons Learned along the Way. In A.S. Lee, J. Liebenau, and J.I. DeGross (Eds.), *Information Systems and Qualitative Research* (pp. 225-245). London: Chapman & Hall.

Trauth, E.M. and Jessup, L. (2000). Understanding Computer-mediated Discussions: Positivist and Interpretive Analyses of Group Support System Use. *MIS Quarterly*, Special Issue on Intensive Research, 24(1), 43-79.

Walsham, G. (2000). Globalization and IT: Agenda for Research. In R. Baskerville, J. Stage and J.I. DeGross (Eds.), *Organizational and Social Perspectives on Information Technology* (pp. 195-210). Kluwer Academic Publisher, Dordrecht, The Netherlands.

ENDNOTES

1 See Trauth (2000, "Appendix: Notes on Methodology", p. 402-403) for a discussion of this.

2 For example, I once received reviewer feedback on a conference paper that was limited to 12 pages. Given this space constraint the emphasis of the paper was on presentation of the results not description of the methodology. Yet a reviewer noted as a weakness that we did not sufficiently explain the perspective of the interpretive researchers who were conducting the research. Had we done so, there would have been little space for the results of our work.

About the Authors

Richard L. Baskerville is Associate Professor of Information Systems and Chairman of the Department of Computer Information Systems, College of Business Administration, Georgia State University. His research specializes in security of information systems, methods of information systems design and development, and the interaction of information systems and organizations. His interests in methods extends to qualitative research methods. Baskerville is the author of *Designing Information Systems Security* (J. Wiley) and many articles in scholarly journals, practitioner magazines, and edited books. He is an associate editor of *The Information Systems Journal* and *MIS Quarterly*, and a member of the editorial boards of *The European Journal of Information Systems* and *The Information Resources Management Journal*. Baskerville's practical and consulting experience includes advanced information system designs for the U.S. Defense and Energy Departments. He is former chair of the IFIP Working Group 8.2, a Chartered Engineer under the British Engineering Council, a member of The British Computer Society and Certified Computer Professional by the Institute for Certification of Computer Professionals. Baskerville holds degrees from the University of Maryland (B.S. *summa cum laude*, Management), and the London School of Economics, University of London (M.Sc., Analysis, Design and Management of Information Systems, Ph.D., Systems Analysis).

Dubravka Cecez-Kecmanovic earned her BS in Electrical Engineering at the University of Sarajevo, her MS in System Sciences and Information Systems at the University of Belgrade and her PhD in Computer Science and Information Systems at the University of

Ljubljana. Until 1992 she had been with the Informatics Department, Faculty of Electrical Engineering, University of Sarajevo. Since arriving in Australia in 1993 she has held the positions of Professor and Head of School of Information Systems and Management Science, and Deputy Dean of the Faculty of Commerce and Administration, Griffith University, Brisbane; and Pro-Vice-Chancellor Research and Consultancy, University of Western Sydney, Hawkesbury. She is currently Professor and Founding Chair in Information Systems at the College of Law and Business, University of Western Sydney. She is also a founder and a leader of the Research Group Information Systems - Knowledge Management in Organisations (IS-KOMO), where she established three major research programs: Social Impacts of Computer-Mediated Communications, Web-based Learning and Teaching and Knowledge Management Enabling Environment funded by Government grants and industry. She has published in the field of social systems of information and government information systems, information systems development methodologies, groupware technologies and more recently in Web-enhanced cooperative learning and teaching, computer-supported organisational knowledge sharing and management, computer-mediated communications, and critical IS research.

Heinz K. Klein earned his Ph.D. at the University of Munich and in 1998 was awarded an honorary doctorate from the University of Oulu, Finland. He is currently working at the MIS Department of Temple University in Philadelphia. Well known for his contributions to the philosophical foundations of information systems research and methodologies of information systems development, he has published work on rationality concepts in ISD, the emancipatory ideal in ISD, emergent systems, alternative approaches to information systems development and principles for conducting and evaluating interpretive field studies. His articles have appeared in the best journals of the field such as the *CACM, MISQ, Information Systems Research*, and others. He received the 1999 MISQ Best Paper Award. In addition, he has contributed to some widely quoted research monographs in IS as author, co-author or editor and served on several editorial boards.

Allen S. Lee is Eminent Scholar and Professor of Information Systems at Virginia Commonwealth University, in the United States. Lee has been an *MIS Quarterly* editor since 1990 and is its current editor-in-chief. He publishes in the areas of research methods, case studies, and electronic communications. With Jonathan Liebenau and Janice I. DeGross, he co-edited the book, *Information Systems and Qualitative Research* (London and New York: Chapman & Hall, 1997). He and M. Lynne Markus served as senior editors of the special issue of *MIS Quarterly* on "Intensive Research in Information Systems: Using Qualitative, Interpretive, and Case Methods to Study Information Technology." Lee has been a proponent of the integration of different research methods within the same research stream and, if page limitations permit, within the same study.

Throughout her distinguished career, **Professor Enid Mumford** has explored socio-technical problems in systems and work organization design. She conducted groundbreaking ethnographic research on industrial relations in both the Liverpool docks and the North West coal industry. In doing so, she pioneered the concept of participative design of systems. Professor Mumford has published widely on all aspects of systems design. A particular contribution to the field has been her work on the involvement of users in the design task. Her methodology takes both technical and human factors into account in the design process and attempts to arrive at a solution that takes into account both efficiency and job-satisfaction needs. Enid Mumford is Emeritus Professor of Manchester Business School where she was a Professor of Organizational Behavior. She was also director of the MBA program. She received the American Warnier Prize for contributions to information science, an honorary doctorate from the University of Jyvaskyla, Finland, and a Leo award from the International Conference on Information Systems (ICIS) and the Association for Information Systems (AIS).

Michael D. Myers is Professor of Information Systems in the Department of Management Science and Information Systems at the University of Auckland, New Zealand. His research interests are in the areas of information systems development, qualitative research methods in information systems, and the social and organizational aspects of information technology. His research articles have been published in journals such as *Accounting, Management and Information Technologies, Communications of the ACM, Ethics and Behavior, Information Systems Journal, Information Technology & People, Journal of Information Technology, Journal of International Information Management, Journal of Management Information Systems, Journal of Strategic Information Systems, MIS Quarterly,* and *MISQ Discovery*. He is co-author of three books. He currently serves as senior editor of *MIS Quarterly*, editor of the *University of Auckland Business Review*, associate editor of *Information Systems Journal* and *Information Systems Research,* and as editor of the *ISWorld Section on Qualitative Research*.

Steve Sawyer is an Associate Professor at The Pennsylvania State University's School of Information Sciences and Technology, where he conducts social informatics research. His current work encompasses the social processes of software development, systems implementation and related organizational changes. Sawyer earned his doctorate at Boston University and has also served on the faculty of Syracuse University's School of Information Studies. To date, he has published in journals such as *Computer Personnel, Communications of the ACM, IBM Systems Journal,* and *Information Technology and People*. With co-authors Rob Kling, Holly Crawford, Howard Rosenbaum and Suzie Weisband, his first book, *Information Technologies in Human Contexts: Learning from Social and Organizational Informatics* was published in 2000.

Ulrike Schultze is Assistant Professor in Information Technology and Operations Management at Southern Methodist University. Her research focuses on knowledge work, particularly informing practices, i.e., the social processes of creating and using information in organizations. Dr. Schultze has written on hard and soft information genres, information overload, knowledge management and knowledge workers' informing practices. Her more recent projects are in the areas of electronic commerce. Her research has been published in *MIS Quarterly*, *The Journal of Organizational Computing and Electronic Commerce* and *The Journal of Strategic Information Systems*. She received her bachelor's and master's degree in Management Information Systems from the University of the Witwatersrand in Johannesburg, South Africa, and her Ph.D. in MIS from Case Western Reserve University, Cleveland, Ohio.

Eileen M. Trauth is Professor of Management Information Systems in the College of Business Administration at Northeastern University in Boston. Her research interests are at the intersection of socio-cultural and organizational influences on IT and the IT profession. She has just completed a multi-year investigation of socio-cultural influences on Ireland's information economy. Her book, *The Culture of an Information Economy: Influences and Impacts in the Republic of Ireland* was published in 2000. Trauth has recently embarked upon a multi-country study of socio-cultural influences on gender in the IT profession. She has also published papers on global informatics, information policy, information management, IT skills, and qualitative research methodology. In addition to editing this book, Dr. Trauth is co-author of *Information Literacy: An Introduction to Information Systems*. She has taught and conducted research in Australia, Canada, France, Ireland, The Netherlands, New Zealand and the UK, and serves on the editorial boards of several international journals. Dr. Trauth received her Ph.D. in information science from the University of Pittsburgh.

Cathy Urquhart is Senior Lecturer and Head of Information Systems at the University of the Sunshine Coast, Queensland, Australia. Her research interests are in analyst-client communication and the social processes of adoption of IT including eCommerce. She also teaches qualitative research methods for IS and takes a strong interest in how they are applied in the field. She has published a number of qualitative research papers nationally and internationally in conferences and journals including *ICIS, IFIP 8.2* refereed proceedings, and *Information Technology and People*. She was the recipient of the "Outstanding Paper" Award for *Information Technology and People* in 1999. She is an enthusiastic member of her national and international IS community, and is involved in various editing and reviewing activities, and regularly reviews qualitative research manuscripts for IS journals. She holds a PhD in Information Systems from the University of Tasmania, Australia.

Eleanor Wynn received a PhD in linguistic anthropology at the University of California, Berkeley. Her 1979 dissertation was the first ethnographic study directed at issues in computerization. Her work was sponsored by Xerox Palo Alto Research Center, where she worked for four years. Following that, Dr. Wynn worked for Bell Northern Research and as a senior scientist, and then as a consultant to major corporations in the computer and telecommunications industry, until her present position as knowledge mapping manager for Intel Corporation. Her work has covered a broad range of technologies and their relevance to workplace competencies held in groups. She has applied theories from social science and phenomenology to practical problems in the assessment of technology in the organization. Dr. Wynn is editor-in-chief of *Information Technology & People*, an information systems journal dedicated to leading-edge concepts and research in information technology worldwide.

Index

A

abstraction risk 207
access to information 93
accessibility 148
action research 47, 62, 69, 193, 273
actor network theory 34, 39
advocacy 32
ambiguity 195
analyst-client interaction 110
analytic 22
analyze 194
authenticity 126
automation of work 81
axial coding 116

B

behavior-centric researchers 245
behavioral-technological interactions 244
business environment 87

C

case study 179
categories 105
change 46
chaos and complexity studies 36
chaos theory 23
classification 220
client/server computing 173
coding 104, 277

coding paradigm 128
cohesion of the states 35
collaborative research 200
communication role 68
communicative action 144
communicative practices 151
competitive intelligent analysts 93
Completion Risk 206
computer supported cooperative work 228
computer-based systems 55
computer-supported cooperative 5
computing infrastructure 165
confessional writing 85
constant comparison 107
conversation analysis 228
covert strategic action 154
critical ethnography 79, 152
critical IS research 143
critical research 143, 273
critical social theory 142
critical theory 142, 160
cultural critique 85

D

data collection 165
data reduction 180
database designers 198
decentralised structures 49
design and development 81
design group 67